Visions of the Cosmos

Ever since the introduction of digital imaging and data-gathering, the technologies for observing our universe have advanced in huge leaps. Today, astronomers enjoy the use of a great range of telescopes and sensing instruments, both on the ground and in space, from which they study almost everything in the cosmos. In 1995, the authors first published *Hubble Vision*, about the science of the Hubble Space Telescope. It caused a sensation and became an international best-seller, and was followed by a highly successful second edition in 1998. *Visions of the Cosmos* is the next logical step, and presents the universe through the eyes of the world's great astronomy installations: the Hubble Space Telescope, the Chandra X-Ray Observatory, the Very Large Telescope, the twin Keck Observatories, and the many other telescopes and solar system probes in use today.

This spectacularly illustrated book is a comprehensive visual exploration of astronomy. It features the latest and most stunning images, and provides a magnificent portrayal of the beauty of the cosmos. The accompanying text is an accessible guide to the science behind the wonders, with clear explanations of all the major themes in astronomy. Readers will learn of the remarkable discoveries being made about our solar system, the stars, nebulae, galaxies, and the structure of the universe. It also presents a look at some exciting observatories of the future. *Visions of the Cosmos* offers a fascinating view of the universe that will appeal to anyone with an interest in the stars.

Carolyn Collins Petersen is a science journalist and creator of educational materials for astronomy.

John C. Brandt is an adjunct Professor of Physics and Astronomy at the University of New Mexico.

The Tarantula Nebula (in the Large Magellanic Cloud) as seen through the Wide-Field Imager camera at the European Southern Observatory in Chile. The intricately sculpted cloud of gas and dust blazes with the light of massive young stars.

Visions
of the Cosmos

Carolyn Collins Petersen

Loch Ness Productions

John C. Brandt

University of New Mexico

CAMBRIDGE
UNIVERSITY PRESS

PUBLISHED BY THE PRESS SYNDICATE OF THE UNIVERSITY OF CAMBRIDGE
The Pitt Building, Trumpington Street, Cambridge, United Kingdom

CAMBRIDGE UNIVERSITY PRESS
The Edinburgh Building, Cambridge CB2 2RU, UK
40 West 20th Street, New York, NY 10011–4211, USA
477 Williamstown Road, Port Melbourne, VIC 3207, Australia
Ruiz de Alarcón 13, 28014 Madrid, Spain
Dock House, The Waterfront, Cape Town 8001, South Africa

http://www.cambridge.org

First published 2003

Printed in the United Kingdom at the University Press, Cambridge

Typeface Quadraat 11/15 pt.   System LaTeX 2_ε [TB]

A catalog record for this book is available from the British Library

ISBN 0 521 81898 2 hardback

Contents

Preface

Each time we look at the sky, we get a glimpse of a universe undergoing the inexorable process of change. Planets roll along in their orbits, stars are born, live, and die, and galaxies wheel through the cosmos. When we chance to glance upon these objects, we catch them for a moment in time – frozen in the act of evolution.

Visions of the Cosmos is a multi-wavelength snapshot album that shows how the universe looks. Appreciating beautiful photographs gives us all a chance to marvel at objects in the cosmos, to think about how they formed, what they're made of, and what contribution they make in the grand scheme of things. When it comes right down to it, looking at a picture inspires a flight of fantasy, a momentary mental visit to a distant planet or star cluster. Those imaginary trips across space and time generate questions about how such objects formed, what they're doing, and what will happen to them in the future. Throughout this book, we've tried to give you a flavor of how astronomy works so that you can understand what these visions of the cosmos have to say. Each image has something to say – and we invite you to ponder what you see, because the visions that the cosmos inspire are an important part of your appreciation of it.

In chapters 1 and 2 we present a general introduction to the science and tools of astronomy. From there, we move out through the solar system in chapter 3, visiting some well-known landmarks along the way. Beyond the solar system lie the stars, and chapter 4 concentrates largely on the processes of star birth and star death. Chapter 5 deals with galaxies, while chapter 6 discusses cosmic change – the evolution of the universe from the Big Bang to the present. We end our tale at chapter 7 with a brief look at the tools future astronomers will use to study the cosmos.

Our motivation in writing this work was simple: use pretty, sometimes provocative images to take readers on a tour of the universe. The good news is that there are millions of pictures out there from which to select. The bad news is that there will never be enough room in any book to present everything you ever wanted to see in the universe, so we had to be selective. We hope we've come up with an enjoyable way for you to explore the planets, stars, and galaxies.

Acknowledgments

Humans are immersed in a datastorm of information pouring down on us from the sky every second. It's a torrent of astronomy data, and the best that researchers can do is stand underneath it and use a few buckets to gather in the data. Observatory "data buckets" collectively form a giant, multi-wavelength eye on the sky. They are scattered across and around the Earth, and at a few other planets just for good measure. This assemblage of ground-based and space-borne platforms continually funnels the results of the astronomical datastorm to thousands of observers around the world.

This book would not be possible without the results and interpretations of those astronomy researchers – both professional and amateur – who constantly dip their buckets into the cosmic data stream and try to make sense of what they've collected. Their work comes to the rest of us through papers, books, and images. We appreciate the access and professional courtesy we received from all of our colleagues in the profession as we completed this work. In particular we want to thank amateur observers Thomas Dobbins and Pedro Ré, astronomers Fred Espenak (NASA Goddard Space Flight Center), Alex Filippenko and Weidong Li (University of California at Berkeley), Robert Hurt (Infrared Processing Astronomy Center), Taku Ishida (Kamioka Observatory), Rhian Jones and Dinesh Loomba (University of New Mexico), Colin Lonsdale (MIT Haystack Observatory), David Malin (David Malin Images), Ian Morrison (Jodrell Bank Observatory), Hermann Mikuz (Crni Vrh Observatory), Seth Redfield (University of Colorado), and Allan Treiman (Lunar and Planetary Institute). We are very much in the debt of the observatory public information offices, institutes, space agencies, and universities, who make a vast quantity of astronomical knowledge available through their World Wide Web sites. In particular, David Finley (National Radio Astronomy Observatory), Peter Michaud (Gemini Observatory), and John Stoke (Space Telescope Science Institute) went out of their way to provide updated materials and access to us.

We extend special appreciation to David Tytell of Sky Publishing Corporation, who provided immensely helpful editorial direction, and to Mark C. Petersen of Loch Ness Productions for commenting on the final manuscript, as well as his Photoshop® expertise in preparing many of the book's images. Tom Hess (University of New Mexico) was invaluable in assisting us to transfer large manuscript and image files across the country. We also want to thank those who commented on various drafts of the chapters: Daniel Altschuler of the Arecibo Observatory, Roger Cappallo of the MIT Haystack Observatory, and writing colleague and astronomy enthusiast Gregory Redfern. We also thank Madeleine Needles of the MIT Haystack Observatory for her invaluable assistance in literature searches. Grateful thanks go to our patient friends at Cambridge University Press – Jacqueline Garget, Anna Hodson, Jayne Aldhouse, and Simon Mitton.

We are also indebted to our families, whose encouragement and understanding made the process of data mining and book writing worthwhile. To Dorothy Brandt and Mark C. Petersen, we give our thanks and our love.

Finally, we thank you – the reader – for choosing this book. The universe is a strange and wonderful place, and it's a privilege to have you along as we explore a few of its most beautiful sights.

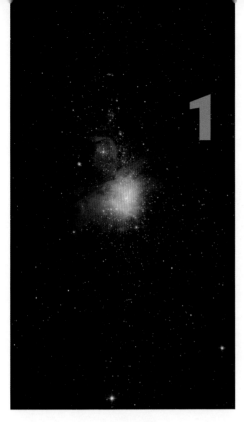

Eyes on the sky

1

The sky is the daily bread of the eyes.
Ralph Waldo Emerson

For my part I know nothing with any certainty but the sight of the stars makes me dream.
Vincent Van Gogh

Stargazing is a practice as old as human thought and forms the foundation of one of the oldest sciences. Modern sky-watchers are descended from generations of astronomers who used the skies for various ends throughout history: to worship, rule, and eventually as a place to learn more about the universe. As it did for our ancestors, the cosmos dangles tantalizing visions before our eyes and dares us to come out and explore. Then, preposterously, it challenges us with seemingly impassable distances. Here we sit, embedded in a universe full of amazing things that we can see but cannot touch or smell or taste. Fortunately, we are endowed with an innate sense of curiosity and the magnificent intellect to figure out those things that are currently beyond our physical grasp.

The complexity of the cosmos is reflected everywhere we look, from the microscopic world of the atom to the macrocosmic domain of the galaxies. How can we possibly appreciate such a breathtaking range of space and time? At first glance it seems impossible to measure and understand things we can't possibly hope to experience first hand. We have to start somewhere, so we classify and measure the cosmos in every objective way we can: points and blobs of light become planets, moons, comets, asteroids, stars, nebulae, galaxies, and superclusters.

We have a tougher time with the gulfs of space that separate us from these objects. Clearly they aren't the friendly Earth-based distances we're used to, but we can make a stab at comprehending them anyway. First we build up a stepladder of distances familiar to us, beginning with the short distances we travel to work or school. Travel to another country takes us farther away. The leap from Earth's surface takes us to the Moon, a major step for humans of any age. The outer limits of the solar system lie even farther away. We cast inquisitive eyes from our little planetary niche out to the stars in our own galaxy. Beyond that we look to other galaxies wheeling through the cosmos in clusters. Just at the faintest edge of our

Figure 1.1. The Earth and the Moon from the Galileo spacecraft as it sped away toward Jupiter in 1992. Seen from this distance, our watery blue world looks serene and welcoming.

Figure 1.2. Mars continues to fascinate
sky-watchers. Today we monitor it from Earth orbit
using everything from amateur instruments to the
Hubble Space Telescope. A fleet of robot explorers
continues to map its surface and study its
landforms. Human exploration of the Red Planet is
a cherished dream, but our first step onto the
Martian surface remains many years away.

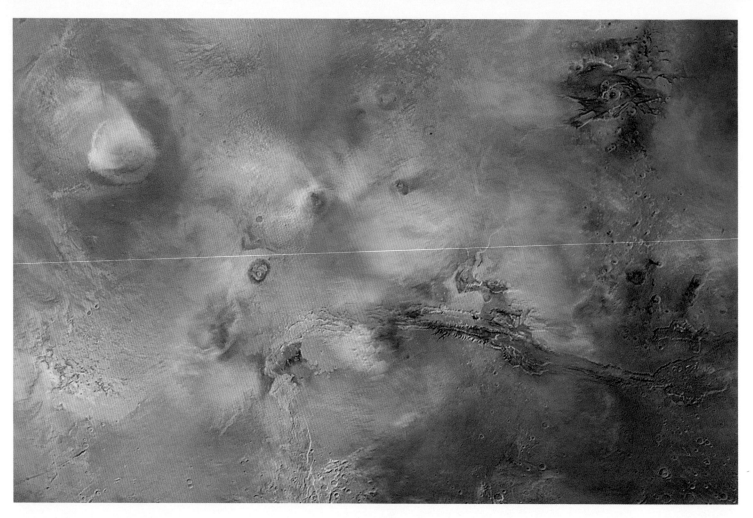

Figure 1.3. Mountains and valleys are an integral part of the Martian landscape. In this Mars Global Surveyor flyover image, Olympus Mons – the largest volcano in the solar system – lies wreathed in clouds in the upper-left corner, flanked by its companions. At the lower right a system of canyons splits the crust for thousands of kilometers. On Earth the Martian "grand canyon" would span the North American continent.

astronomical eyesight – so far away that their light traveled billions of years to reach us, we find faint glimmerings that point to the birth of the universe.

The same distances that hinder our immediate exploration compel us to define them in terms that encompass many thousands of millions of our familiar, comfortable kilometers. This is where units like light-year, parsec, and megaparsec come into play. These terms take the unimaginable and make it somehow easier to swallow. The reality of astronomy is that we grow comfortable with these terms – and each new discovery gives us more to learn. Soon we are tossing out comments about the spiral structure of a galaxy a few billion light-years away as if it was simply in another country.

To understand astronomical distances, we need to start right here on Earth. From our home we look out and see the Sun, Moon, and planets and wonder how far away they are. Celestial mechanics, simple observations, and in some cases, the use of radar signals help determine distances within the solar system. So, for example, the distance from Earth to the Sun is 149 million km. Light, traveling at 300 000 km per second, takes 8 minutes to traverse that distance, making the Sun 8 light-minutes away. The moon is 380 000 km from Earth, and light takes 1.27 seconds reach its surface. So you could say that the moon is 1.27 light-seconds away. Mars is around 228 million km from the Sun, or 13 light-minutes away. The distant outer planet Pluto is 5900 million km from the Sun, or 5.5 light-hours away.

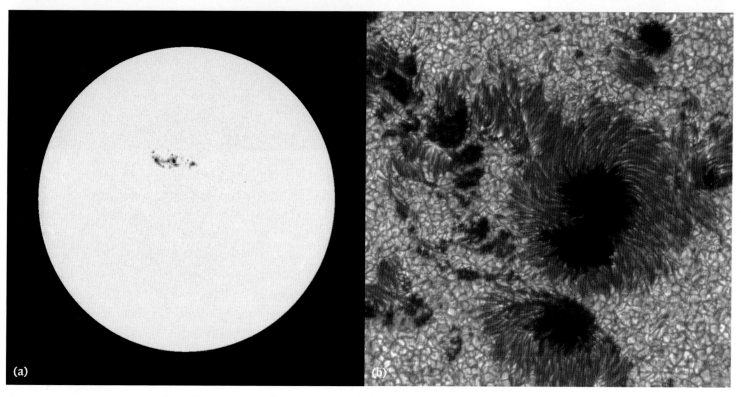

(a)

(b)

Figure 1.4. **(Top)** Saturn's most enduring feature is its complex and beautiful ring system, stretching across hundreds of thousands of kilometers above the planet. Storms come and go in its hydrogen atmosphere, hinting at dynamic changes beneath the cloud tops.

Figure 1.5. The Sun is the major source of heat and light in the solar system, and also provides astronomers with their closest example of a star. **(a)** The Sun's visible surface is often mottled with cool, dark regions called sunspots. As the Sun rotates they move across its face. **(b)** A close-up ground-based view of a July 15, 2002 sunspot group reveals granular mottling on the Sun's surface and dark penumbral filaments surrounding the darkest part of the spot. Our growing understanding of the Sun's cyclic blemishes may well help probe the workings of other stars that exhibit similar spots.

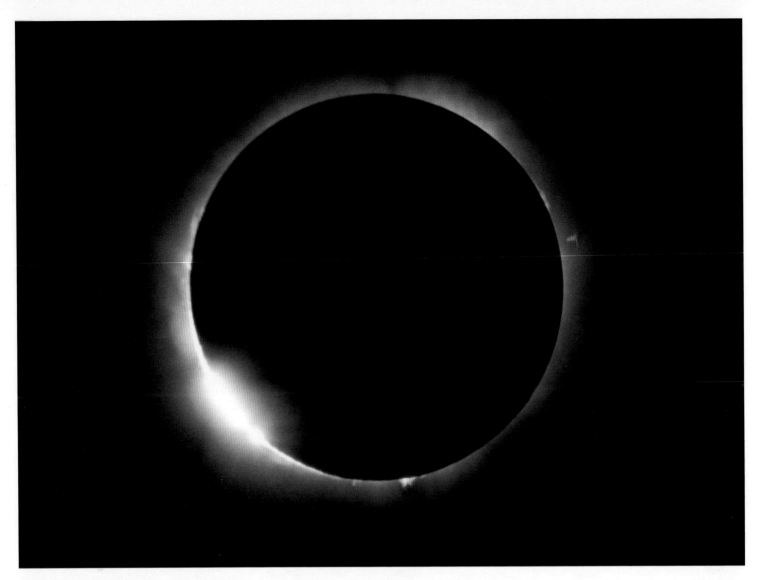

Figure 1.6. Ancients feared total eclipses, but today's observers travel great distances to experience "totality." During these rare occurrences, we catch glimpses of the corona – the Sun's superheated outer atmosphere. The solar wind originates here, streams past the planets, and out to interstellar space.

These distances are relatively easy to manage, but when we start to look at nearby stars and galaxies the numbers get huge. Recitations of numbers like 4100 billion km can get pretty clumsy. So, astronomers have developed some distance units that are easier to comprehend. First there's the parsec – which is 206 265 astronomical units, or even easier, 3.26 light-years. Using these units the distance to the nearest star – Proxima Centauri – is 1.3 parsecs. The Large Magellanic Cloud – a companion to our Milky Way – is 52 000 parsecs away or 52 kiloparsecs. For even larger distances, the system once again gets too big to handle, so we speak in terms of millions of parsecs, or megaparsecs. No matter how you

Figure 1.7. **(Opposite)** Star birth is one of the last great mysteries of stellar life. The Orion Nebula – some 1500 light-years away – is the closest star nursery in our galactic neighborhood. The nebula is a bubble of heated gas associated with the giant Orion Molecular Cloud. The clouds of gas and dust surrounding Orion's star-birth regions hide much star-forming action from our view. Wavelengths of light longer than our eyes perceive can pass through the dusty environment and allow astronomers a window onto the stellar nursery. These infrared wavelengths are captured by the instruments in the 2 Micron All-Sky Survey (2MASS). The data are translated into false-color images that simulate what an observer with infrared-sensitive eyes might see. The blue and red colors are assigned to light emitted by the various gases in the Orion clouds.

Figure 1.8. Continuing an infrared zoom into the heart of the Orion Nebula. The European Southern Observatory in Chile zeroed in on the Trapezium cluster at the heart of the Orion Nebula. This star-forming region is home to hundreds of hot, bright young stars easily seen in optical images. Astronomers search for the existence of brown dwarfs, which stand out at infrared wavelengths. These objects are too small and cool to be stars, yet too warm to be planets.

Table 1.1 *A comparison of distances*

Distance unit	Kilometers[a]	Light travel time
Light-second	3×10^5	1 second
Light-minute	1.8×10^7	1 minute
Astronomical unit (AU)	1.5×10^8	8 minutes
Light-year	9.5×10^{12}	1 year
Parsec (pc)	3.1×10^{13}	3.3 years
Kiloparsec (kpc)	3.1×10^{16}	3.3 thousand years
Megaparsec (Mpc)	3.1×10^{19}	3.3 million years

[a]Powers of ten (3×10^2) is a shorthand way of referring to large numbers. Multiply 10×10 to get 10^2, which equals 100, and then multiply that by 3 to get 300. Thus, 3×10^5 equals $10 \times 10 \times 10 \times 10 \times 10 \times 3 = 300\,000$. (See Glossary for a complete explanation of powers of ten notation.)

measure them, though, these are huge numbers describing incredible distances (Table 1.1).

Distances, of course, aren't the only gulfs we must bridge in our search for an understanding of the cosmos. There is an intellectual gap between seeing an image of a distant object, and recognizing what it is and how it formed. A picture

Figure 1.9. The Hubble Space Telescope joined the hunt for brown dwarfs in the Trapezium region of the Orion Nebula. Using a special camera sensitive to infrared light, it spied 50 of these substellar babies (the faintest objects in the image) among the more than 300 newborn stars hidden in this fiery nursery.

(a)

(b)

Figure 1.10. Star-birth crêches are littered throughout the galaxy. Contrasting views of the famous "Pillars of Creation" in the Eagle Nebula (M16) show the value of multi-wavelength approaches in the search for star birth. **(a)** The Hubble Space Telescope focused on a region of space where interstellar gas and dust clouds are being eaten away by ultraviolet radiation from a nearby star. **(b)** The Infrared Spectrometer and Array Camera (ISAAC) on the European Southern Observatory's Very Large Telescope can penetrate through the clouds and revealed nests of stellar formation and several very young, relatively massive stars in the tips of two of the pillars.

Figure 1.11. The seeds of stars are contained in regions of hydrogen gas mixed with interstellar dust that astronomers call cold molecular clouds. The Hubble Space Telescope peered at this set in the constellation Centaurus, nicknamed "Thackeray's Globules." They float in space, silhouetted against a more diaphanous gas cloud that is itself lit up by radiation pouring from nearby young stars more massive than our Sun.

Figure 1.12. The famed Horsehead Nebula is a cold molecular cloud protruding into another region of heated hydrogen gas. At the top of the horse's head a bright rim separates the cloud from the surrounding hydrogen gas region. This is an ionization front where hot photons from nearby stars are destroying the dust and heating up the gas. In response, the heated hydrogen gas emits light – creating a glowing emission nebula.

of a crater in Arizona tells us something about an event in our planet's past. On the surface of a planet like Mars, we find evidence of a watery past, completely different from what exists today. Jupiter's clouds roil and writhe before our eyes. Out beyond the solar system, we see immense clouds of gas and dust that we now know are the birthplace of stars – but astronomers a hundred years ago didn't know what to make of them. Elsewhere an ominous-looking nebula is all that remains of a supergiant star that died in a violent cataclysm. A picture of a warped galaxy invokes our curiosity about how it formed and how it got the way it is. As we look farther out in space, we see older and older objects which, because of the time it takes for their light to travel to us, looking younger than they are at the present time. As we observe and categorize the objects we observe, we ask a great many questions: What are they? How did they form? How will they die? What is their place in the evolution of the universe?

Astronomy concerns itself with answering these questions and many others. It would be exciting if astronomers could explore the cosmos from the bridge of some spacefaring vessel that was not bound by any of the laws of physics. Then it would only be a matter of traveling across the galaxy faster than light speed and examining interstellar mysteries with highly advanced long-range scanners and advanced probes. Unfortunately, experiencing the evolution of stars and galaxies on any short-term scale is impossible for us because we are a short-lived species

Figure 1.13. Understanding how stars die is as essential as knowing the circumstances of their births. One of the most often-asked questions astronomers hear is "What will happen to the Sun?" As it nears its end, the Sun will shed its atmosphere and collapse to a slowly cooling white dwarf. The light from the shrunken Sun will light up the shells of gas it blasted away earlier in its death throes. Here, NGC 6751 shows how the Sun might look in its own planetary nebula stage.

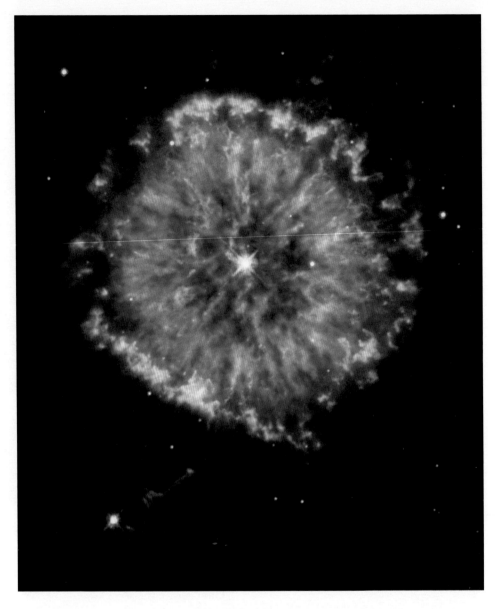

and we don't have faster-than-light ships. Even if we did, the lifespan of even several generations of astronomers wouldn't be enough to sample more than a brief moment in the ever-changing life of a star or the incredibly long existence of a galaxy. But all is not lost, because the universe throws us a lifeline of sorts. Everywhere we look in the cosmos, we find planets, stars, and galaxies at many different points in their evolution. So, while we don't have the ships to take us to those objects, we have created long-range sensors and space-based probes to help us study the universe from our Earth-bound point of view. The answers these tools give us depend on something we take very much for granted – light.

Astronomy and light

Think of light as a kind of astronomical Rosetta Stone. Like the ancient hand-carved rock that contained clues to understanding early Egyptian hieroglyphic writing, a beam of light is our guide to comprehending the

Figure 1.14. The most spectacular type of star death occurs when a massive star explodes in a supernova. The filamentary structures seen in this Hubble Space Telescope image are the remains of a massive star that exploded some 160 000 years ago in the Large Magellanic Cloud. This seemingly placid-looking structure hides a spinning neutron star – the likely remnant of the original supergiant.

Figure 1.15. The largest stellar cities are called galaxies. These collections often number hundreds of billions of stars, and fill the universe with astonishing forms and shapes. **(a)** The pinwheel shape of M74 is a typical open spiral. It lies about 35 million light-years away from us. **(b)** Ring-shaped galaxies like Hoag's Object boast circular regions of star birth surrounding a nucleus of old stars, and likely formed as the result of a close encounter of two galaxies. **(c)** The upper left spiral arm of NGC 6872 has wispy blue star-forming regions, a sure sign that it has recently interacted with another galaxy – perhaps the neighboring IC 4970. **(d)** The greenish, sombrero-like galaxy – called ESO 510-13 – has a warped equatorial dust lane that appears to be a recent addition. A galactic merger in the distant past may also explain its odd appearance.

Figure 1.16. Nearly everywhere we point a telescope we can see a cosmic backdrop alive with galaxies – out as far as our detectors can see. This Hubble Deep Field – named for the Hubble Space Telescope and the lengthy set of observations that produced the image – contains millions of galaxies of nearly every shape and size.

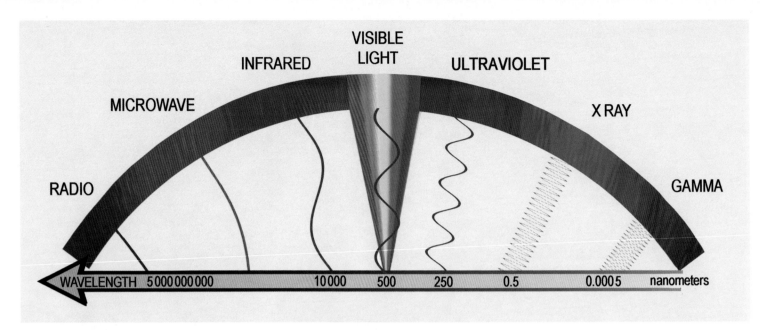

RADIO MICROWAVE INFRARED VISIBLE LIGHT ULTRAVIOLET X RAY GAMMA

WAVELENGTH 5 000 000 000   10 000   500   250   0.5   0.0005   nanometers

Figure 1.17. The spectrum is a continuum of electromagnetic waves, arranged according to frequency and wavelength (given here in metric units of nanometers).

complexity of the universe. Light contains an incredible amount of information about the objects that radiate and reflect it. Want to know about a star's temperature and chemical makeup? Its velocity and direction as it travels through space? Its rotation rate and magnetic field strength? Does it dance in a mutual orbit with another star? Is there a planetary system? To find out, simply study all the wavelengths of light it radiates.

In one sense, the history of astronomy can be characterized as the history of humans learning how to decode the mysteries of light. Once we figured out that light was the key to the universe, we developed a vast array of devices to capture light and analyze it in all its forms. But, how does light encode such a large amount of information?

Light is funny stuff. Depending on how and when you study light, it can be treated as either a wave or a particle called a photon. Think of a photon as a packet of energy. Each one has an energy value, given in joules (1 joule is 1 watt of power radiated for 1 second). You could also think of the photon as a piece of a light wave, which carries energy across space.

What would it look like if you could slow a light wave down? Imagine throwing a pebble into a pond. You watch a series of circles moving outward from the point where the stone hit the water. Each circle is a wave with a peak and a low point. The wavelength is the distance between each peak, and frequency is the number of peaks that pass by as you watch. Light moves in a similar way. Light has a speed (of 300 000 km per second), a wavelength (usually given in units of metric distance), and a frequency (given in units of cycles per second, called hertz). When you turn on a 100-watt light bulb, it emits about 10^{20} photons every second and dissipates the rest of its energy as heat. The light from the bulb radiates at frequencies between 400 and 670 terahertz. An astronomer would describe the same light as radiating between 450–750 nanometers.

Light falls into a spectrum of wavelengths called the electromagnetic spectrum. The EMS encompasses everything from gamma rays, x rays, and ultraviolet light through visible wavelengths, infrared, and radio waves. To put it into perspective, imagine if all the wavelengths of the electromagnetic spectrum could be spread out across an area about the length of an American football field (91.4 m). The light we see with our eyes (visible light) would be in a band only 0.0000003 m wide – much narrower than one of the blades of grass on that imaginary football field. All of visual astronomy has been done from that narrow wavelength region – and gives you an idea of how much more there is to learn when we look at the cosmos in other wavelengths.

To form some idea of the nature of light, imagine going outdoors on a warm summer's day. As you sunbathe, photons of visible light enter your eyes and enable you to see your surroundings. If a rainstorm has recently passed by, you see different wavelengths of light in the colors of a rainbow. You feel infrared wavelengths as warmth from the Sun, and some ultraviolet radiation burns, or tans, your skin. The Sun also emits higher-energy ultraviolet and gamma radiation, but you do not experience it as you lie there on the sand because Earth's atmosphere screens most of it out.

Now turn that experience to stargazing. Everything in the universe puts out radiation, often in many wavelengths. Human eyes are limited to one set of wavelengths – the optical, or visible, range. As you scan the sky with your naked eye or telescope, you observe objects that radiate in the visible wavelengths of light. If the Moon or planets catch your eye, you are seeing the visible light they reflect. They don't generate their own visible light but reflect light from the Sun. Stars and galaxies provide you with radiated visible light because they *do* generate their own photons.

If you had multi-wavelength eyes, the universe would look very different, and you'd be able to see things in – well – a different light! Begin at the radio end of the spectrum, where the wavelengths range from 1000 m to 1 cm. Your radio receivers would put you in tune with the central regions of galaxies where strong radio signals belie the existence of jets of matter streaming away from the site of something massive and energetic. The Crab Nebula's radio image would be brightest around its pulsar – a tiny but massive neutron star spinning 30 times a second on its axis. You'd also be able to detect Jupiter's radio signals, emanating from the energized particles that give off strong signals as they spiral around lines of force in the planet's magnetic field. Ironically, although these frenetic activities generate energetic signals, by the time they reach your radio sensor, they would be barely more than a cosmic whisper.

If you were sensitive to the subset of radio signals called microwaves (wavelengths of 1 mm to a few centimeters), molecular clouds (like the ones that exist in the Orion Nebula) would look especially bright to you. You'd also have the ability to "see" the slowly cooling echoes of the creation of the universe as a sort of mottled, low-color backdrop to the stars and galaxies. Again, as with radio, you would need to strain to detect these faint signals.

Infrared eyes would let you see energy given off by objects in the universe that are at room temperature or slightly warmer. They radiate at wavelengths ranging from micrometers to centimeters. What would you see? Cool stars (often called

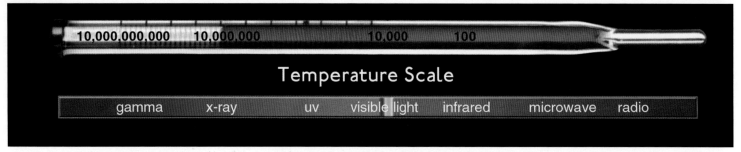

Temperature Scale

10,000,000,000   10,000,000   10,000   100

| gamma | x-ray | uv | visible light | infrared | microwave | radio |

Figure 1.18. The electromagnetic spectrum can also be arranged by temperature (in degrees Kelvin) – showing the range of wavelengths from those associated with energetic, high-temperature objects and processes to the coolest and least energetic.

brown dwarfs or sub-stellar objects), planets, clouds of particles between the stars, interstellar molecules, nebulae, and so-called infrared galaxies would populate your skies.

Visible-light eyes – which are what you have now – are sensitive to light emitted or reflected from objects at wavelengths between 400 and 700 nm. They open up the richness of the sky to you, but at a price. Dust prevents you from peering deeply into the hearts of objects or from sensing the most energetic things in the cosmos.

Eyes sensitive to ultraviolet wavelengths (between 10 and 400 nm) would let you distinguish the elements present in an interstellar gas cloud, measure the density and temperature of gas clouds, and watch as the radiation from hot young stars ate away at the clouds that gave them birth.

The most energetic processes in the universe give off the most energetic radiation: x rays and gamma rays. These are very short wavelengths – smaller than 10 nm. X-ray eyes have long been a staple of hokey science-fiction stories or flashy ads for "x-ray glasses" that would allow you to look through people's clothes and see their bones. But with true x-ray eyes, you would only see the most energetic things in the universe. Many of the most energetic things in the universe pour out radiation in the form of x rays. To an x-ray detector, the universe is a violent, ever-changing place of million-degree temperatures, rapidly accelerating objects, and incredibly strong magnetic fields that can threaten such stable objects as stars. Your sky would include hot gas clouds, the cores of galaxies, the regions around black holes, and oddly enough – cometary comas that emit x rays as they plow through the highly electrically charged solar wind streaming from the Sun.

The most energetic processes in the universe give off radiation called gamma rays. These wavelengths are extremely short (less than about 0.1 nm) and extremely energetic. The gamma-ray view would be dominated by supernova explosions, the areas around black holes, and something mysterious called gamma-ray bursters.

In a sense, we already do have multi-wavelength researchers out there. They're astronomers (both professional and amateur) who supply us with the multi-wavelength eyes that nature didn't give to humans. These researchers have devised ways to study every regime of light, using both ground-based and orbital instruments.

Astronomy: The observational science

Observation is the foundation of astronomy, and the first step in the scientific method – a framework for scientific research that covers everything from anatomy to zoology. Once someone has seen an object or event or process, the next step is to describe what has been observed. The observer then formulates a hypothesis to explain the object or event. This is where things get interesting, particularly in astronomy. As soon as the observer forms a hypothesis, the door is opened to many other possibilities that may just as easily explain the observation. Ideally, once a hypothesis has been suggested, other astronomers should be able to use it to predict similar events or observations at a later time. The researcher gets to test the hypothesis by conducting the observations again. The success of the observation will test the validity of the prediction, and the hypothesis behind it. Every astronomical discovery – from the determination of how the Moon got its craters to the latest theories about the age of the universe – depends on this rigorous method of inquiry for success.

Most astronomical observations can be divided into three categories: *imaging*, *photometry*, and *spectroscopy*. Radio observations are collected by sensitive dishes similar to satellite television dishes but much larger.

In *imaging*, a signal is recorded on film or on a glass plate or captured by an electronic camera called a charge-coupled device (CCD), which contains collecting "buckets" similar to the imaging "chips" now found in many consumer cameras and camcorders. A CCD is a specialized detector containing an extremely sensitive chip that stores light in a miniaturized series of "bit buckets" called "picture elements" or pixels. The collected information is transferred from the chip to a computer for storage and analysis, and ultimately is processed into an image. A conceptually similar process is used to combine data from an array of radio telescopes to from a radio "image."

Sometimes a telescope is pointed at a field crowded with objects, such as a star cluster. If we are studying just one star in the cluster, it helps if the instrument can separate one star from another. This is the concept of *spatial* resolution – the ability to distinguish objects in a crowded field from each other and to produce clearly defined images of them.

Some objects in the universe change extremely rapidly, and scientists learn more if they can capture as much change as possible. Certainly, cameras attached to telescopes can image something many times over the course of an observation, or for a much longer time than the human eye can gaze at an object, but there are limits to just how many "pictures" can be taken with ordinary film cameras, plates, and single CCDs. So, specialized cameras and detectors are used to capture an event as it happens over a period of time. This is the concept behind high *temporal* (time) resolution, and the best systems will give us many "snapshots" of an event within fractions of a second.

Photometry is the practice of measuring the intensity of light from an object, and it often depends on instruments that can record with high temporal resolution. A photometer can be thought of as a very sensitive light meter, similar to that used in flash photography. Photometers are used to determine fluctuations in light intensity from a variable star, for example. The determination of intensity of light from stars and other objects in the universe is extremely important because from those studies come the determination of luminosity (brightness), or magnitudes. Quite often an astronomer will refer to an 8th-magnitude star or a 5th-magnitude comet. What the magnitude number refers to is the relative brightness of the object compared to other objects. The brighter stars have smaller magnitude numbers, while dimmer stars have larger numbers. Sirius is the brightest star in the night sky, with a visual magnitude of -1.5; Canopus is -0.7, while Betelgeuse is around 0.5. Under ideal conditions, the dimmest stars we can see with the naked eye range around 5th or 6th magnitude. In the solar system, the Sun shines at magnitude -27, while distant Pluto is 16.8. Some faint, distant galaxies shine at magnitude 30 or dimmer!

Different photometers measure different wavelengths, so there are whole sets of instruments sensitive to infrared and ultraviolet light as well as to visible light. Photometry is a very systematic way to classify celestial objects and to observe how their changes of light output affect the way they appear to us.

Spectroscopy takes light and divides it into its component wavelengths. We are all familiar with some everyday types of spectroscopy: white light shining through a prism and sunlight shining through raindrops to form a colorful rainbow. These demonstrate very simply that the Sun's light (or light from any white-light source) radiates all the wavelengths in the spectrum.

Spectroscopy helps us answer a variety of questions that might not be answered if we relied solely on undispersed light to study objects. What is the chemical composition of an interstellar gas cloud? How hot is a star? What gases and ices are contained in a comet? How fast is a jet of material streaming from the center of a galaxy? The methods of spectroscopy allow us to answer these questions by studying the way in which stars and galaxies, comets, and planets emit and absorb light.

The simplest spectrographic tool is the spectroscope, which breaks light down into very fine divisions by wavelength. A prism will work, but most modern astronomical spectrographs – instruments that collect light, break it down and record the results on film or as computer data – use diffraction gratings. These are glass plates or mirrors scored with very thin lines. As the light shines across the grating, it is broken up into very, very fine wavelength divisions. You can begin to see how a diffraction grating works by looking at a compact disc in sunlight. A very fine, continuous rainbow of colors appears.

Spectroscopy is a very powerful tool in chemistry, where the identifying characteristics of an element can be determined with great precision. It is a fairly simple process in the lab – you apply heat to an element and study the light given off as it burns. Each element or substance (molecules of gases, for example) has a very distinctive "fingerprint," or spectrum. Generally, a spectrum looks like a smooth continuum of color, broken by very bright or dark lines. The information that spectroscopy reveals about an object is encoded in these lines. The concept of *spectral* resolution describes the ability to cleanly separate adjacent features in a spectrum.

The basic rules of spectroscopy were formulated in the nineteenth century by a German chemist named Gustav Kirchhoff. His laws of spectral analysis describe what sort of spectra appear as elements are burned. Kirchhoff's first law states that a hot, high-density gas or an incandescent solid body will radiate a continuous spectrum.

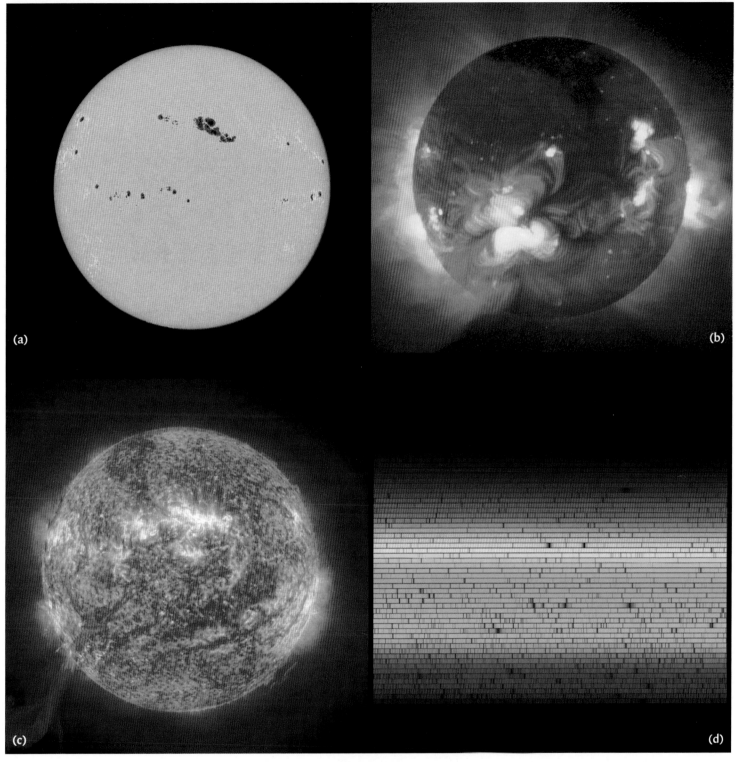

Figure 1.19. A multi-wavelength view of the Sun illustrates different visions our star presents in visible, x-ray, and extreme ultraviolet wavelengths (a, b, and c respectively). The flare extending out from the Sun in the lower left image stretches out nearly 40 Earth diameters. A spectrum of the Sun breaks sunlight into small wavelength ranges and shows the chemical fingerprints of the elements in the solar atmosphere (d).

Figure 1.20. Astronomy data is often charted on x–y graphs. This one plots radiation intensity over an hour's time during a solar flare that occurred on March 6, 1989, recorded by two satellites that recorded x-ray bursts from the Sun. The data are from the GOES weather satellite and the Solar Maximum Mission, which studied energetic radiation released during solar flares. The plot begins just before the flare breaks out, traces the high output of radiation during the flare, and then graphs the gradual fade-out of the signal as the flare subsides. Three wavelength "regimes" are plotted here in a stack. The intensity scale of the radiation is about 1–100 counts per second per square centimeter for the gamma rays; 1–1000 counts per square centimeter for the hard x rays; and 10^{-6} to 10^{-3} watts for soft x rays.

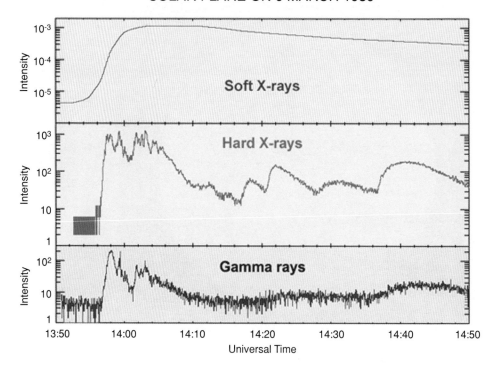

The Sun is a good example of this principle. That bright, yellow star we see each day is quite a different-looking beast in other wavelengths. If we could look at it with x-ray or ultraviolet eyes, we'd see radiation at those wavelengths as well.

The second law states that a hot, low-density gas will produce what is called an "emission-line" spectrum, i.e., elements that are abundant in the object will appear as very bright lines in its spectrum. Kirchhoff's third law states that when a source of continuous radiation, such as a star, is viewed through a cooler, low-density gas, an absorption-line spectrum will be produced. The study of absorption spectra is a particularly ingenious way of determining what lies between us and a star. The spectrum of that star will show dark lines where certain wavelengths of light have been absorbed by clouds of material in interstellar space. All an astronomer needs to do is to compare that spectrum to laboratory-defined spectra of elements suspected to exist in the star to identify its makeup. Absorption and emission spectra of the Sun indicate that our parent star is rich in a variety of chemical elements.

Astronomers take the output of their detectors and create images or graphs to help in their analysis of the object under study. We've alluded to astronomical detectors being huge "light buckets" but they are also huge "bit buckets" as well. All the data coming from the instruments goes through some amount of data processing – and in astronomy, there's a tremendous amount of data generated. An important part of processing all that information is the ability to visualize it. We all respond to "pretty pictures" of planets and galaxies and those are important to study in astronomy. A number of data-processing tools help astronomers turn their information into images for further study – even if the original observations were at wavelengths invisible to the human eye. In these cases color information is applied to the "ones and zeroes" to give us a simulation of what an object would look like if we could see it in its native wavelength. In addition to the many visual light images in this book, we also present a number of infrared and ultraviolet "images" that were produced in this way. Another way to study astronomical data is to take the information and plot it on a graph. It may not look particularly photogenic, but such plots are immensely important to researchers and often convey as much exciting information as a picture does.

Treasures from the temples

So, what do astronomers watch? The combined output of our observatories gives us a view of the universe unlike anything our earliest ancestors (and even some of our closer ones) could imagine. Our view of the universe changes daily and with each new detector we build or mission we send. While we have by no means explored the entirety of the cosmos, the continual flow of information from the sky helps us synthesize an understanding of what we see. From the nearest planets to the limits of the observable universe, the findings have been remarkable.

Our interest in the solar system lies in our need to understand the origin and evolution of Earth. The beauty and diversity of our sister worlds, the asteroids, comets, and many moons of the solar system, tell us much about the conditions that existed during the formation of the solar system. Each planet is a unique world, with environments that range from the familiar to the completely alien (when compared to our home planet).

The other planets in the solar system are undergoing constant change. We have evidence of this from the atmospheric and surface processes that continually alter their faces with volcanism or winds or the movement of ice and water. Because of these changes, planets require constant monitoring if we are ever to understand them completely. For example, a worldwide network of amateur and professional astronomers continually observes the planet Mars, watching for any changes in its atmosphere or surface markings. In 2001, observers kept the planet under surveillance, watching for a series of "Mars flashes" that had been predicted to occur when the Sun, Earth, and Mars were at a favorable geometry. The best explanation for the flashes was sunlight glinting off of a thin layer of ice overlaying sand dunes in the crater Schiaparelli. The flashes did occur and observers around the world shared notes about what they saw. A few weeks later, astronomers watched the onset of a storm that began in late June and shrouded the planet for several months. Coupled with ground-based and space-based observations of the Red Planet, observations like this give a more complete picture of the changes that occur at Mars.

Not all observing is done to detect changes on faraway objects. Earth is at the mercy of a large collection of as-yet undetected small bodies that pose a continual threat of collision with our planet. It's a threat that some governments are starting to take seriously. NASA and other agencies are spearheading the development of more sensitive detectors and automated sky surveys to search the skies for early-approach warnings of asteroids and comets.

Far beyond the planets (and our concerns for asteroidal Armageddon) lie the stars – more than 200 billion of them in our home Milky Way galaxy alone. Billions

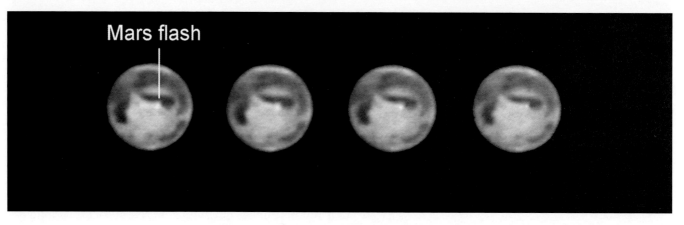

Mars flash

Figure 1.21. In the early morning hours of June 7, 2001 a group of Mars observers comprising both amateur and professional astronomers observed a series of surface brightenings in a region called Edom Promontorium, near the Martian equator. This sequence of video screen shots shows white flashes just to the upper right of center. The remarkable thing about this set of images is that it was made while observers were watching the glints from Mars in real time from a distance of about 70 million km through a variety of small telescopes!

more stars populate countless numbers of galaxies stretching out as far as we can see. Wherever we look in a starfield we are likely to see stars at all stages of their lives. The detectors we use help to decipher the structure of a star's life from the evidence we see before us. In some nebulae we are uncovering the shrouded mysteries of star birth. In others we measure the sometimes-explosive events signaling the end of a star's brilliant life.

Star life – those millions of years between formation and death – forms the basis for much of astronomy's lure. Some stars travel the universe alone, but many more move through space in pairs, triplets, and larger clusters. Some have what look like planetary systems in orbit around them. Stellar characteristics – rotation rates, chemical makeup, velocity, age, color, and temperature – all are fair game for the astronomer's study.

The galaxies in which stars exist give us clues about the formation of the universe, the density of its many populations, and perhaps even a glimpse into the far future of our universe. Galaxies harbor star birth regions, black holes, and sites of star death – and studying those processes and objects in other places gives us important clues about these same occurrences in our own galaxy.

Astronomers also seek to answer those questions for which our answers are, at best, incomplete. Although we classify things in the universe into planets, stars, galaxies, and quasars – at what seems to be ever-increasing distances away from us – we have yet to find the distance to the "edge" of the universe. In reality there is no such boundary – but rather a beginning – the moment of creation when this universe and its potential for galaxies, stars, planets – and life – were formed. Unfortunately this moment is shrouded from our sight.

But, not to worry! As the strength and sensitivity of our detectors increases, astronomers at the "cutting edge" of science are cataloging the shapes of the earliest galaxies and measuring wavering signals from a time some 300 000–400 000 years after the Big Bang. They're looking at old familiar faces in different wavelengths of light, discovering amazing things in the process. Far from being a "done deal" astronomy is a science still on the first steps of an amazing journey through space and time.

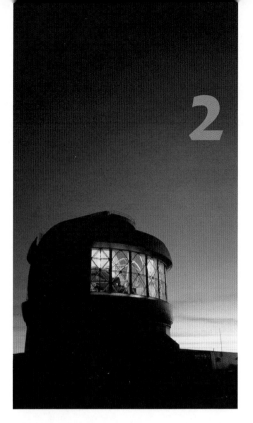

2

Telescopes: Multi-frequency time machines

Telescopes are in some ways like time machines. They reveal galaxies so far away that their light has taken billions of years to reach us. We in astronomy have an advantage in studying the universe, in that we can actually see the past.

Sir Martin Rees, Astronomer Royal of Great Britain

Good luck, Mr. Hubble

John Grunsfeld

Observatories: Temples for sky-watchers

There is no question about it – observatories *feel* like mystical places. In a sense, to people who are not familiar with telescope operations and technology, they *do* seem to perform a bit of magic by bringing the universe to our doorstep. And they have always done so, even when their builders didn't quite understand what they were seeing.

When we explore the history of astronomy, we take a journey backward in time and we see the world through the eyes of our earliest stargazing ancestors. The *really* ancient astronomy was probably practiced by anyone who had the time and energy to stay up late after a full day of hunting, sheep-herding, warmongering, or whatever else it was that early humans did during the daylight hours. What sparked that first interest in the sky? Was it the daily miracle of sunrise? When did someone finally figure out that the Moon kept appearing in a regular rhythm and decide to chart it along with the Sun's slow trips across the sky? What did those early skygazers make of the stars as they watched them rise and set against familiar landscape features?

The earliest known observatories are little more than stone constructs like Stonehenge in England or the stone sanctuaries on the island of Menorca, or open-air temples carved into mountainsides like the Intihuatana at Machu Picchu, Peru. These present the tangible proof of an interest in the sky by observers who drew sophisticated conclusions about the motions of the sky and seasonal changes affecting the surface of the Earth.

In principle, today's observatories aren't much different from the earliest sky-watching pavilions. To be sure, they are more widespread than in earlier times and stuffed with complex machines. Installations dot the landscape of every continent on Earth and swing through space in the near-Earth and

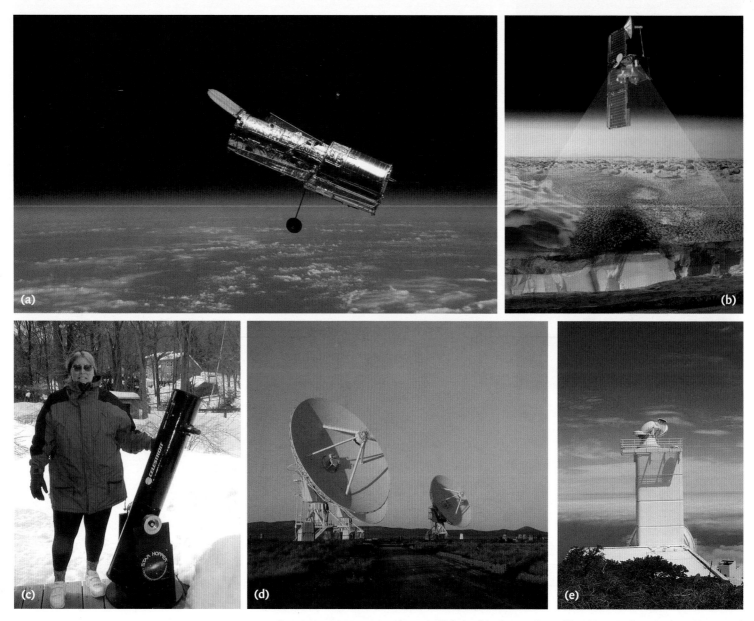

Figure 2.1. Astronomers seek out the light in all its forms, given off by objects in the universe, with some of the most fantastic tools ever invented: telescopes. Observatories can be found everywhere – in our backyards, sprawled across dusty plains, perched atop high mountains, and even in space. They are, quite literally, our eyes on the sky. The range of telescopes and observatories in use today is quite remarkable. **(a)** The Hubble Space Telescope does its observing from low-Earth orbit, while planetary explorers like Mars Odyssey **(b)** orbit other worlds and return data to waiting scientists. Ground-based observatories range from small amateur scopes **(c)**, to large radio arrays like the Very Large Array in New Mexico **(d)**, and specialized facilities like the Swedish 1-m solar telescope at La Palma **(e)**.

Figure 2.2. **(Opposite)** Optical telescopes traditionally have featured mirrored surfaces – like the 8.1-m Frederick C. Gillette Gemini telescope atop Mauna Kea, Hawaii **(a)** – to collect and focus light to instrument arrays. To get around the problems posed by ever-larger mirrors, a new generation of facilities is utilizing flexible, segmented mirrors that can be "figured" with actuators. **(b)** The Hobby–Eberly Telescope at McDonald Observatory in Texas has 91 segments forming an 11 × 10-m hexagon. This gives the telescope a 77.6 square meter light-collecting area.

(a)

(b)

Figure 2.3. The multi-wavelength Milky Way. To understand the different wavelength regimes that astronomers observe in, imagine how the Milky Way would look if we could "see" all the radiation it emits and reflects. From top to bottom: in visual wavelengths, our home galaxy – the Milky Way – is riddled with bright glowing clouds of gas and dust, interspersed with dark dust clouds that absorb light. To gain an infrared view of our Milky Way, astronomers mapped the sky at infrared wavelengths around 2 μm as part of the 2-Micron All-Sky Survey (2MASS). This image contains data from nearly 100 million stars going down to magnitude 13.5. To a person equipped with radio eyes, the Milky Way would look like this map generated by the Very Large Array. The plane of the galaxy is quite evident, although no individual stars stand out. The luminous sources near the plane are pulsars, regions of star birth, and the remnants of supernova explosions. The mission of the Extreme Ultraviolet Explorer (EUVE) spacecraft resulted in an all-sky ultraviolet map. The stripes in the image are the result of the spacecraft's orbital scanning technique. The constellation Orion, with its three distinctive belt stars appears near the left of center. The x-ray Milky Way, as seen by the Röntgen Satellite (ROSAT). The brightest spots indicate regions of intense activity that result in increased x-ray emissions. The Compton Gamma-Ray Observatory was sensitive to the most energetic radiation in the universe. This all-sky survey measured hot spots in the Milky Way galaxy. It was the first of its kind, performed by the Energetic Gamma-Ray Experiment Telescope (EGRET) on the satellite.

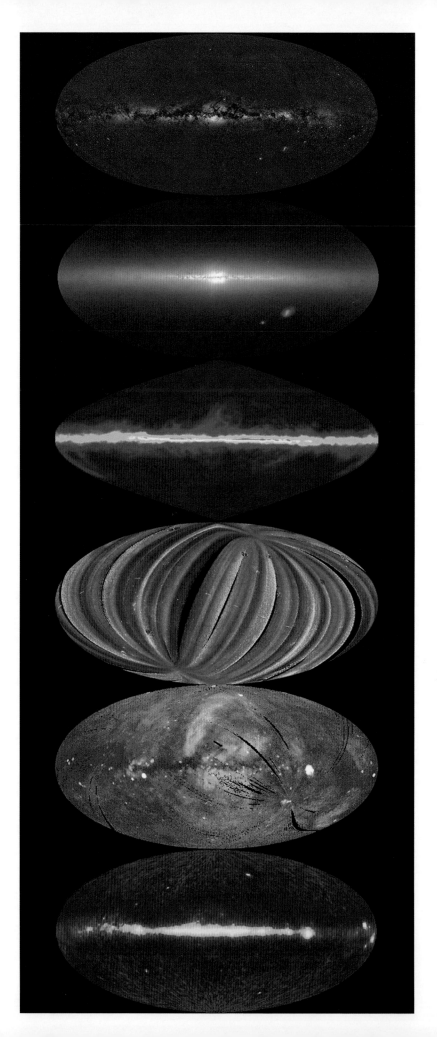

outer-solar-system environments. At latest count, more than 300 professional astronomy installations and thousands more amateur facilities round out the complement of Earth-bound observatories.

Since the early 1960s our eyes in the sky have included sub-orbital rocket flights, balloons, orbital satellites, and fly by spacecraft. These include a panoply of observatories regularly written about in the media, like the Hubble Space Telescope (HST), the Chandra X-Ray Observatory, the Solar Heliospheric Observer (SOHO), the Infra-Red Astronomical Satellite (IRAS) of the early 1980s, the late 1990s ROSAT (Röntgen Satellite) mission, the Hipparcos astrometry satellite, and many others.

Beyond the reach of Earth's gravity we have sent dozens of probes (and 12 humans) to the Moon. The Mariner spacecraft, the Magellan orbiter, the Giotto probe to Comet Halley, the Mars, Vikings, Pathfinder, Global Surveyor, and Mars Odyssey missions have explored the other worlds of the inner solar system. Following in their trails are the Stardust comet mission and the Near-Earth Asteroid Rendezvous Probe. The glories of the outer solar system were revealed to us (but have barely been explored) by the prodigious stream of data flowing from the Pioneer, Voyager, and Galileo spacecraft. The Cassini probe is among the latest to venture out to Saturn on a multi-year mission of exploration.

Stargazers transported from our distant past to a mountaintop observatory such as Australia's Anglo-Australian telescope or the launch of a space-based observatory would be stunned at the intricate instrumentation astronomers use to study the cosmos. However, the principles behind today's complex arrays of detectors are actually quite simple: gather as much light as possible from objects in the universe. What sort of light astronomers gather depends on what they're using to gather it. Unfortunately, there is no instrument that senses all wavelengths of light perfectly. So astronomers use different detectors to study different wavelength regimes.

A telescope is the most familiar astronomy tool around. At its heart, it is a device that gathers electromagnetic radiation in the form of ultraviolet, visible, infrared, or radio wavelengths. In many telescopes, a mirror focuses and reflects light to other instruments, or sensors such as film cameras, radio receivers, and CCD cameras. From the images produced by these cameras, astronomers can learn about such things as brightness of an object, its shape, location, and relationship in space to other objects.

The bigger the telescope you build, the more light (or other signal) you can gather. However, any telescope mirror has limits on how big it can be. A large mirror can gather more light, giving its users a chance to look at dimmer and/or more distant objects. However, mirrors cannot be made infinitely large because the structures to support them would have to be infinitely large as well. Then there is the budget, which becomes a problem long before you hit the size limit. The shape of a mirror is crucial to its ability to concentrate incoming light to a tight focus, and large parabolic or curved mirrors bend and sag under their own weight. Astronomers are getting around these limitations by proposing and building new generations of lightweight and segmented mirrors, as well as those whose surface can be adjusted in real time. In some cases, they are linking together many smaller

Figure 2.4. The European Southern Observatory consortium operates two major observing sites in Chile. (a) One is the Very Large Telescope (VLT) array which is located on Cerro Paranal and has a collection of 8.2-m and 1.8-m installations. (b) To get an idea of how large an 8.2-m. telescope is, notice the person standing beneath the main optical assembly at the VLT's Kueyen telescope. (c) The La Silla Observatory is made up of a series of optical telescopes ranging in size up to 3.6 m. The Swedish–ESO Submillimeter Telescope is also operating at this site.

Figure 2.5. The Anglo-Australian Observatory at Siding Spring, Australia. The AAO operates two optical telescopes – the 3.9-m Anglo-Australian Telescope (upper right) and the 1.2-m UK Schmidt telescope (lower left).

Figure 2.6. The observatories atop Kitt Peak, near Tucson, Arizona. The 4-m Mayall Telescope (far right) is the largest optical instrument on the mountain. Just downhill is the University of Arizona's 2.3-m telescope, plus the 0.9-m Spacewatch instrument, whose function is to find and study near-Earth asteroids. Spacewatch has a new 1.8-m telescope built after this picture was taken. At the far left, is the 3.5-m WIYN telescope. Wisconsin, Indiana, and Yale Universities funded it, with contributions from the National Optical Astronomy Observatories. Other domes on the mountain include a visitors' telescope, a 0.9-m telescope, a 2.4-m telescope, the Burrell Schmidt installation, and the automated facility belonging to the Southwestern Association for Research in Astronomy. The triangular-shaped building is the McMath–Pierce Solar facility, the world's largest solar telescope. In front of the McMath–Pierce telescope is the Vacuum telescope, which contains a giant vacuum chamber running almost the full height of the building.

(a)

(b)

Figure 2.7. Observatories in Paradise. The 4200-m-high Mauna Kea volcano in Hawaii houses the world's largest collection of observatories for optical, infrared, and submillimeter astronomy. **(a)** There are currently 13 facilities here, including the twin Keck domes, the University of Hawaii telescopes, the Canada–France–Hawaii installation, the Smithsonian Submillimeter Array, the United Kingdom Infrared Telescope, and the NASA Infrared Telescope Facility. **(b)** The National Astronomical Observatory of Japan maintains the 8-m optical/infrared Subaru facility at Mauna Kea, while the United Kingdom, Canada, and the Netherlands collaborate in the operation of the James Clerk Maxwell Submillimeter telescope **(c)**. The northern installation of the twin Gemini telescopes **(d)** is one of the latest observatories built on Mauna Kea. Its southern counterpart **(e) (facing page)** is in Chile.

(c)

(d)

(e)

Figure 2.8. Sacramento Peak at Cloudcroft, New Mexico is home to the towering white Richard B. Dunn Solar Telescope, and the neighboring Evans Solar Facility coronagraph installation.

Figure 2.9. Quite possibly the coldest observatory on Earth inhabits a cold dry plain at the Amundsen–Scott South Pole station. The Antarctic Submillimeter Telescope Observatory is operated year round, focusing on spectroscopic observations of atomic and molecular clouds in galaxies.

Figure 2.10. Radio telescopes are a good example of how big a radiation "bucket" can get. Some of the more spectacular examples include (a) the 300-m-wide "dish" at Arecibo, Puerto Rico, and (b) the 100-m Robert C. Byrd Green Bank Telescope in West Virginia. (c) To understand the scale of the Green Bank dish, the figures in the image are construction workers on one portion of the dish as it was being welded together. (d) The Lovell radio telescope at Jodrell Bank, England. Radio astronomy is often done using arrays, where a series of telescopes are linked together to make a larger observing "eye." (e) The Very Large Array west of Socorro, New Mexico is a good example of such an array.

Figure 2.11. **(a)** Observatories that don't look the part. The Cosmic Background Imager was built specifically to study the cosmic microwave background radiation left over from the early universe. It is located at an altitude of 5080 m in the Chilean Andes. **(b)** It might be hard to think of a big underground tank full of fluid as an astronomical instrument, but the Super-Kamiokande detector is really a very specialized neutrino "telescope." Neutrinos are particles emitted by our Sun, supernova explosions, and the centers of active galaxies. Researchers used Super-Kamiokande and the neutrino detector in the Homestake Gold Mine in Lead, South Dakota to make fundamental discoveries about the creation of neutrinos in the Sun and were awarded the 2002 Nobel Prize for physics.

(a)

(b)

Figure 2.12. The observatories and telescopes belonging to Portuguese amateur astronomer Pedro Ré. By day Ré works as a marine biologist at the University of Lisbon, but by night he turns his collection of instruments to the sky to observe everything from the Moon and planets to distant deep-sky objects.

detectors spread out across large areas, and combining the incoming signals, electronically forming a giant "eye" on the sky.

As we discussed in chapter 1, the astronomy we're most familiar with – the sort where we step outside and look up at the stars and galaxies – makes use of visible light. It's what our eyes evolved to see. Optical astronomy operates in a narrow band in the electromagnetic spectrum that stretches from the near infrared to the near ultraviolet. Astronomers actually measure this range between 350 and 650 nm – which is nearly the same as our own visual range. The peak sensitivity of our eyes is about 550 nm during the day (which just about matches the wavelength of the peak energy output of the Sun). At night, with dark-adapted eyes, a human's peak sensitivity is about 500 nm.

You can reproduce the visible spectrum by letting sunlight shine through a prism. What you'll get is a spectrum of colors. The wavelengths of light you can't see stretch out far beyond the little slice that you *can* see. Take radio astronomy as an example. To detect the emissions that come our way in the radio and microwave regions of the electromagnetic spectrum, astronomers use large dish-shaped metal reflectors that gather the faint radio signals emanating from distant objects, focus and amplify them, and then turn them into a signal format that can be manipulated to make a radio "image." Radio signals can be collected by one dish or by a series of linked dishes in a process called "interferometry" (a process that can also be used for optical telescopes). The fainter the object, the more surface area a radio detector must have to observe it. Multiple dishes in an array can

Telescopes: Multi-frequency time machines

combine to give the astronomer more collecting area. Also, the separation of the widely spaced telescopes gives the researcher a very detailed look at very small (or distant) objects.

Most astronomy buffs are familiar with the Very Large Array (VLA) in New Mexico in the United States. To the general public, it is the place in the movie *Contact* where signals from a distant civilization were received. In real life, radio astronomers use its 27 linked receivers to focus on radio-emitting objects as diverse as the planet Jupiter, pulsars, and active galactic nuclei. Occasionally the VLA is linked to more far-flung radio telescopes a continent or more away to gain an even finer "picture" of a radio source. This is where things get interesting because each dish receives the same signal at a slightly different time from the others in the array. Once the signals from all the dishes are combined with appropriate delays and other processing methods, an image can be formed. If this is done properly, the resulting image will be as detailed as if a dish the size of the entire array had collected it, although the sensitivity to faint emission will only be as great as the sum of the areas of the individual telescopes.

All ground-based telescopes (regardless of the wavelengths they detect) are affected by Earth's atmosphere and the growing problem of light pollution. Radio telescopes are particularly sensitive to radio frequency interference and are usually located well away from busy population centers and congested roadways. Atmospheric turbulence causes stars to twinkle, which is very pretty when you're out stargazing with a loved one. But, if you're trying to do precise measurements on an object, that twinkling can ruin an observation run and skew the data. By the same token, light and noise pollution tend to wash out the light and signals from dim, distant objects, also rendering any observations useless.

Ground-based eyes

It should be obvious from the observatories pictured here that astronomers get to do their work from some of the most beautiful and remote places on Earth, and even their own backyards, if they so desire. Many of the world's facilities are located on mountaintops (such as the collections of telescopes in Hawaii and South America and others in Europe, Australia, Japan, and China) and desolate areas far away from city lights. Others, like the Very Large Array near Socorro, New Mexico, or the Penticton radio array in British Columbia, Canada, are located well away from cities in wilderness areas and deserts to avoid the jarring interference of cars, cell phones, and television, and radio transmitters.

While astronomers can go to the ends of the Earth to escape most of the pollution and atmospheric interference, they still end up having to compensate for working under a blanket of gases. They've developed several techniques for dealing with atmospheric effects. One way has been to take many very short observations of an object. This effectively freezes the turbulence produced by the Earth's atmosphere. Then, the individual images are mathematically combined to infer details about the "true" image. This is called "speckle interferometry."

In some wavelengths, astronomers can almost overcome atmospheric effects by adapting the telescope's mirror or other sensors to changes in the atmosphere. The technique is called adaptive optics, and it works like this: the signal from an object is distorted as it passes through the atmosphere. A visible light picture of a star-like point would look blurry and be useless for research. Adaptive optical systems correct for that atmospheric aberration by projecting a laser beam high in the atmosphere to create a synthetic "reference" star. Its light is then analyzed to determine how much the atmosphere is distorting the incoming light. The system uses this information to change the curvature of the adaptive optics mirror very quickly to compensate for the atmospheric distortion in the light, making the reference star appear point-like. The beauty of this procedure is that it can be applied in real time – while observations are in progress! Adaptive optics allows spectroscopy and photometry to be done on short timescales. These systems have allowed their users to virtually remove the effects of the atmosphere from their observations.

Finally, while most astronomers find whatever way they can to get their data from above the atmosphere, several groups of researchers make their way deep underground – to a large holding tank called Superkamiokande lined with detectors and filled with ultra pure water. It is contained in the Kamioka Mine near Tokyo, Japan. Another facility has been in the Homestake Mine in Lead, South Dakota for years – and it's filled with dry-cleaning fluid. What researchers measure with these unique installations are the traces of extremely low-mass particles called neutrinos – which are emitted in mass quantities by the Sun, or during supernova explosions, or possibly from the cores of energetic galaxies. They are difficult to detect directly and there are several approaches. In one, neutrinos whiz through, occasionally colliding with atoms in the fluid. The collisions produce radioactive elements that can be detected by sophisticated instruments.

Eyes in the sky

As we've discussed, the best places to do astronomy are far above Earth's atmosphere with its propensity to swallow up incoming radiation from distant objects or "soak" the view in its own emissions. This is why so many infrared and submillimeter observatories are perched along with their visual counterparts at the highest altitudes on Earth. These are only partial solutions, along with such experiments as the balloon-borne BOOMERanG (Balloon Observations of Millimetric Extragalactic Radiation and Geophysics), and sub-orbital (rocket) experiments. To do astronomy at x-ray, ultraviolet, and gamma-ray wavelengths, however, space-based observing delivers instruments far above our absorbing atmosphere. There have been dozens of observatories in orbit around Earth at one time or another and these satellites – like the Compton Gamma-Ray Observatory and the Hipparcos mission – have done cutting-edge science during their useful lifetimes.

Of the observatories currently in orbit, the Hubble Space Telescope, the Chandra X-Ray Observatory, and NASA's Space Infrared Telescope Facility are good examples of orbiting astronomy platforms. Like other observatories (on orbit and on the ground), they make observations with mirrors and funnel light to an array of science instruments behind the mirror.

The Hubble Space Telescope's detectors are sensitive to wavelengths of light ranging from ultraviolet through visual and into the infrared; astronomers get what they refer to as a high-resolution, multi-wavelength view of the universe. The heart of the telescope is the mirror system – a specially built version of a standard Cassegrain-type telescope. In a Cassegrain telescope, light from an object enters the telescope tube and bounces from a primary mirror to a secondary mirror, which sends the light back through a hole in the primary mirror and into various science instruments.

HST was launched into orbit in April 1990. During its years on orbit, the telescope has had nine different instruments installed at various times. The Wide Field and Planetary Cameras (WF/PCs) and the Faint Object Camera (FOC) give us HST's visual take on the universe. The current WF/PC is sensitive to light ranging from near infrared to the near ultraviolet. The ultraviolet-sensitive Goddard High Resolution Spectrograph (GHRS) and Faint Object Spectrograph (FOS) were used on the telescope from 1990 until 1997. Replacing them were the Near-Infrared

Figure 2.13. The Hubble Space Telescope, named after astronomer Edwin Hubble, is the only observatory designed to be serviced on orbit. Astronauts have visited HST several times to do instrument swap-outs and repairs. It is scheduled to remain in orbit until at least 2010.

Camera and Multi-Object Spectrometer (NICMOS) – an infrared-sensitive instrument, and the Space Telescope Imaging Spectrograph (STIS) – designed to study objects radiating in ultraviolet and visible wavelengths. The High-Speed Photometer (HSP) was HST's "light meter" and was removed in 1992 to make way for a corrective optics system. Although not built as science instruments, the Fine Guidance Sensors (FGSs) serve as star trackers, and also perform astrometry – the science of accurately measuring stellar positions. The Advanced Camera for Surveys (ACS) was installed into HST, replacing the Faint Object Camera. Interestingly, the Faint Object Spectrograph and the backup mirror for HST have found their way into exhibits at the National Air and Space Museum in Washington, D.C.

X-ray eyes on the sky

X-ray astronomy is a relatively recent new science and until recently was little known outside of astronomy circles. It was thrust into the public consciousness in 2002 with the awarding of the Nobel Prize for physics to astronomer Riccardo

Figure 2.14. The Chandra X-Ray Observatory is named for the late astrophysicist Subramanyan Chandrasekhar. Chandra and its European Space Agency counterpart, the X-Ray Multi-Mirror Newton Mission (XMM–Newton), are exploring the x-ray universe.

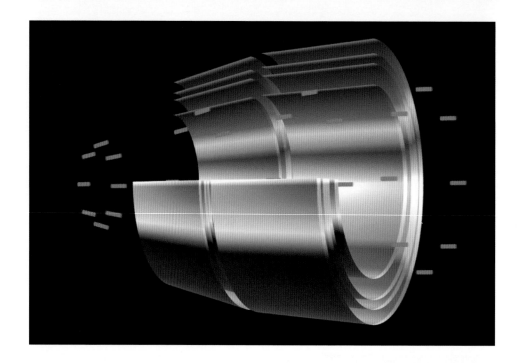

Giacconi for his pioneering work in x-ray research dating back to the early 1960s. Back then, the earliest x-ray detectors rode aloft as rocket and balloon-borne probes to get above the x-ray unfriendly atmosphere. Astronomers have since lofted a number of satellites into orbit that give us an x-ray eye on the sky. Today the best-known missions are NASA's Chandra X-ray Observatory, the XMM-Newton mission launched by the European Space Agency, the BeppoSax mission (developed by the Italian Space Agency with cooperation from the Netherlands Agency for Aerospace Programs), the Japanese Yohkoh solar x-ray mission (which went offline in 2001), and the Rossi X-Ray Timing Explorer (RXTE).

These telescopes are all sensitive to wavelengths of radiation that we normally associate with a visit to the dentist. X rays are produced when gas is heated to millions of degrees by the often violent and extreme conditions found in clouds of hot galactic gases, exploding stars, flare stars, and black holes.

The Chandra X-Ray telescope was launched into a highly elliptical 64-hour-long orbit around Earth in July 1999. The heart of the telescope contains four sets of nested, paraboloid/hyperboloid pairs of "grazing incidence" mirrors. They lie inside a long, tapering tube on the spacecraft. Each highly polished mirror is 83.3 cm long. The x rays enter the spacecraft and glance off the mirror surfaces at a small angle. Then the x-ray signal is recorded by detectors.

The infrared universe

HST and Chandra are part of NASA's Great Observatories program. The Compton Gamma-Ray Observatory (which was decommissioned and allowed to plunge to a fiery end in Earth's atmosphere in 2000) was the third member of the program. The last of these orbiting facilities is called the Space Infrared Telescope Facility (SIRTF). It is designed to detect objects radiating infrared light between 3 and

180 μm in wavelength. This radiation – which emanates from star-forming regions, newly forming planetary systems, the centers of galaxies, and other places where clouds of gas and dust obscure our vision – requires a detector cooled to near absolute zero. SIRTF has a solar shield to shade its three specially cooled science instruments from the Sun. Its heliocentric orbit (trailing Earth) should reduce the effects of heating in the near-Earth environment. At its heart is a 0.85-m telescope, making SIRTF the largest infrared telescope ever launched into space. Its highly sensitive instruments will give us a unique view of the universe and allow astronomers to peer into regions of space that are hidden from optical telescopes.

Traveling with planetary explorers

Although astronomers have been observing the planets from ground-based observatories for centuries, the great age of in situ planetary exploration began when we sent our first robot probes to Earth's moon in 1959. Since then, planetary scientists have sent missions to study nearly everything in the solar system except for distant Pluto. The Sun, planets, a few asteroids, and three comets have all come under the scrutiny of the planetary explorers.

More than a dozen spacecraft and several landers – including some high-profile "lost" missions, have been sent to Mars. The most successful of the early missions were the two Viking orbiter/lander pairs that gave planetary scientists their first high-resolution view of the planet. The next famous (and successful) mission was Mars Pathfinder, with its wandering rover named Sojourner. The most successful recent missions have been the Mars Pathfinder, Mars Global Surveyor, and Mars Odyssey missions.

The search for clues to the evolutionary history of Mars, particularly the fate of its water, has driven scientists to map its surface, measure its magnetic field and atmosphere, and – more recently – map the mineral content of its surface. Mars Global Surveyor (MGS) is a well-equipped explorer. Its camera has lenses that will allow it to capture both wide-angle shots of the planet's surface and images of objects on the surface as small as 1.5 m across. The spacecraft also uses the Mars Orbiter Laser Altimeter – nicknamed MOLA – to bounce beams of light off the Martian surface and back to the spacecraft. The travel time allows scientists to measure the height of various parts of the surface. Mars Global Surveyor's Thermal Emission Spectrometer scans the planet for heat emissions from both the surface and the atmosphere. Mars Global Surveyor sports a magnetometer and electronic reflectometer, giving researchers some measure of the planet's magnetic field, which in turn allows them to make some deductions about the makeup of the Martian interior.

Mars Global Surveyor is also designed to serve as a radio relay station for future surface explorers, and its signals are routinely used to measure changes in the Martian atmosphere. To date, the spacecraft has returned thousands of images of the planet – enough to make up a high-resolution global map of the dry and dusty surface.

Following Mars Global Surveyor's path into Martian orbit in October 2001 is the Mars Odyssey – a mission to map the planet to determine the amount and distribution of chemical elements and minerals that make up the Martian surface. For this purpose, it uses the Thermal Emission Imaging System (THEMIS) and a gamma-ray spectrometer (GRS). As with so many other Mars missions, Odyssey was designed to find out more about the existence of water on Mars. In 2002, the spacecraft's sensors detected the telltale signs of water ice in the upper meter of soil in an area near the Martian South Pole. The mission scientists characterized the find as a layer of dirty ice. Its presence is betrayed by the existence of hydrogen atoms, which in this case indicate water ice. Access to water would be an important factor in future human exploration of Mars. Finally, the spacecraft is equipped with an experiment to study the levels of radiation from the solar wind and cosmic rays at Mars.

The outer solar system has had its fair share of planetary exploration as well. The first visitors to the Jovian worlds were the Pioneer flyby missions in the mid 1970s. Following along their path were the Voyager missions, which lasted from the late 1970s to the late 1980s. Both Voyager 1 and 2 visited Jupiter and Saturn, and Voyager 2 continued on to Uranus and Neptune. All these explorers are on their way out of the solar system, but their missions were wildly successful – giving us our first up-close and personal encounters with the gas giant planets, using instrument loads that included cameras, spectrometers, photopolarimeters (detectors that measure the intensity and polarization of light), and other sensors.

The Galileo probe undertook the longest mission to Jupiter. It was a spacecraft equipped with sensors to study nearly every aspect of the harsh Jovian environment – from its radiation belts to its topmost cloud layers. Galileo completed its mission and plunged into the Jovian clouds in September 2003, but

Figure 2.17. **(Opposite)** Mars Global Surveyor has been orbiting Mars since 1997 and continues to send back images of the Martian surface.

its 8-year mission expanded on the discoveries of the Voyager missions that preceded it. Among the mission's many amazing accomplishments has been a complete map of the volcanic moon Io, the deployment of a probe into Jupiter's clouds, and the discovery of a magnetic field emanating from the moon Ganymede. Galileo also discovered strong evidence that the surface of Jupiter's icy moon Europa is riding atop a slushy saltwater ocean, and that Ganymede and the satellite Callisto may well also have layers of liquid water.

It has been more than two decades since the Voyager spacecraft visited the ringed planet Saturn. Since that time, astronomers have had to content themselves with ground-based observations of the planet and long-distance images taken with the Hubble Space Telescope. The Cassini spacecraft is a joint NASA–European Space Agency mission on its way to a July 2004 arrival at Saturn. It is loaded with a collection of cameras and remote-sensing instruments that will send back a multi-wavelength view of Saturn to waiting scientists. It will be able to sense the properties of Saturn's magnetic field and radiation environment, and sniff out dust particles in orbit around the planet and in its rings. Once there, the spacecraft will release the Huygens probe into the clouds of Saturn's largest moon, Titan. On the way down, the probe will measure the atmosphere and send back images and data of the clouds and surface. Cassini has already begun its exploration of the outer solar system with a flyby of Jupiter and look-ahead observations of Saturn.

Long considered minor parts of the solar system, comets and asteroids are now coming into their own as fascinating objects to study. Stardust is one of the first major missions to a comet in the new millennium. It has already had a close flyby of the asteroid Annefrank in November 2002, on its way to an encounter with Comet Wild 2 in January 2004. Its major goal is to collect dust samples from the comet's tail.

Collecting comet dust is not an easy task, despite the fact that comets scatter the stuff all over space during their orbits. To catch a bit of comet dust, scientists needed something that would capture tiny particles without damaging or destroying them. They came up with aerogel – a silicon-based substance that has a spongy, porous structure. A block of aerogel is mostly empty space and looks like solid blue smoke. Fast-moving particles simply burrow into the aerogel, which absorbs the energy of the impacts and encases the particles for safekeeping. The spacecraft is equipped with an aerogel-filled tray that will trap millions of particles. In 2006, at the end of the mission, the sample return module should parachute safely back to Earth and the waiting laboratories of comet scientists.

Who are the watchers?

More than 11 000 professional astronomers support and use the world's telescope facilities. The community of researchers comprises dozens of disciplines, from solar astrophysics, planetary science, stellar research, high-energy physics, cosmology, and other related sciences like geophysics and astrobiology. In almost all cases, astronomy-related research is funded by government entities, carried out

Figure 2.18. Galileo mission at Jupiter ended when the spacecraft plunged into the clouds of Jupiter after 8 years of mapping the Jovian atmosphere, rings, and moons.

in universities and institutes, and is very team-oriented – employing far-flung groups of people.

A large number of amateur observers also scan the skies both individually and in groups. They use equipment ranging from simple backyard telescopes to sophisticated, fully equipped installations rivaling some of the best university facilities. There are, at best estimate, some 250 000 amateur astronomers in the United States, with perhaps again that many throughout the world. Many more people own telescopes and simply observe the stars, planets, and galaxies from time to time. These sky-watchers come from all walks of life and are united in their love of the stars, planets, and galaxies. As a hobby, pastime, getaway, or whatever you want to call it, amateur astronomy returns humans to our sky-gazing roots, putting us back in touch with the cosmos.

Astronomy itself is a subject with a cosmopolitan appeal and until late in the nineteenth century, amateurs and professionals were simply part of the same (albeit small) community. As the importance of astronomy grew and the costs of observatories exceeded what most individuals could afford, the difference between

Figure 2.19. The Cassini–Huygens mission to Saturn is on its way to a 2004 arrival. Its goal is to study the rings and moons, and send a lander to the surface of the moon Titan. It has already returned images and data from a flyby of Jupiter.

professional and amateur astronomers split the groups. This not-always-amicable divorce was unfortunate because it deprived each side of an important, jointly held heritage.

Today the split is slowly healing. Many exceptional amateurs are making important contributions to the advance of the science in cooperation with a growing number of professionals who see real value in partnering with them. In recent decades a number of groups such as the Association of Lunar and Planetary Observers (ALPO), the American Association of Variable Star Observers (AAVSO), the International Halley Watch (IHW), the Ulysses Comet Watch (UCW), the Center for Backyard Astrophysics, the International Meteor Organization, the International Mars Watch, the International Jupiter Watch, and many others have shared much-needed observational data with their counterparts in professional astronomy.

(a)

Figure 2.20. **(a)** The Stardust mission is on its way to capture the particles shed by a comet as it orbits the Sun. **(b)** A block of aerogel weighs next to nothing, but safely absorbs dust particles and holds them in a collection tray **(c)** for return to Earth.

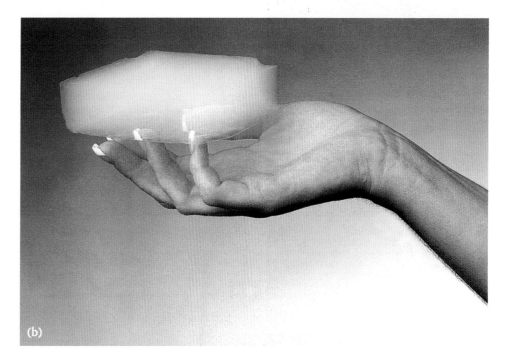

(b)

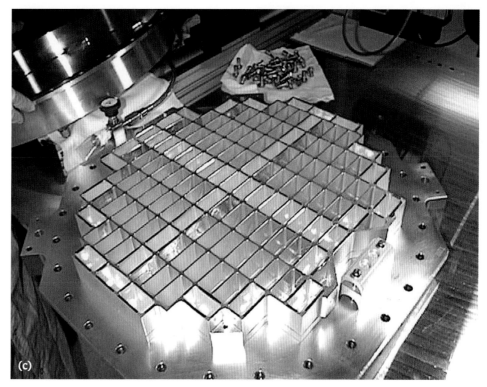

(c)

There is no doubt that twenty-first-century stargazers are being rewarded with some incredible views of the cosmos. Today's observatories and space-based explorers truly extend our eyes to the universe in ways that could only have been imagined even a century ago by the most forward-looking observers. The images in the following chapters are testament to the adaptability and resourcefulness of generations of astronomers determined to get the best possible view of the cosmos possible.

3 Planets on a pixel

It suddenly struck me that that tiny pea, pretty and blue, was the Earth. I put up my thumb and shut one eye, and my thumb blotted out the planet Earth. I didn't feel like a giant. I felt very, very small.

Neil Armstrong, Apollo astronaut

O wad some Power the giftie gie us
To see oursels as ithers see us!

Robert Burns, "To a Louse"

Astronomy is a humbling pursuit. Often it reaches out and confronts us with a reminder that our place in the universe is really pretty small. It's quite sobering to realize that our world and everything on it can be encapsulated in an image less than a pixel wide. Within that tiny bit of space a solar system formed, complete with an inventory of different worlds. And on one fractional bit of that pixel, generations of humans have lived and died.

Yet, small as our cosmic address is, we can't even begin to fathom the deeps of space without knowing something about our own local neighborhood. Until relatively recent times humans had a pretty faulty understanding of Earth's place in the cosmos. Our point of view only improved as our ability to study the solar system changed. In the past century, particularly in the last 50 years, active solar system studies have changed not only our understanding of the other planets, but also of our home world. As a result, now we can now talk knowledgeably about seasonal carbon dioxide levels in the Martian atmosphere, volcanoes on Io, a slushy water ocean underneath the frozen surface of Europa, or compare surface features on distant Uranian moons to ice canyons here on Earth. We routinely focus the Hubble Space Telescope on Jupiter and Saturn to track auroral glows in their upper latitudes, and point x-ray telescopes at comets to understand how a dirty snowball could be the source of such energetic radiation. Scientists confidently plan multi-million-dollar missions to study Titan's cloud-covered surface or Pluto's steadily collapsing atmosphere. Well-equipped amateur astronomers write of seeing flashes on the surface of Mars, track the storms of Jupiter, and routinely scan the sky for asteroids many times fainter than Pluto.

Figure 3.1. What would we look like from a spacecraft cruising by the outskirts of the Solar system? These "blobs in space" are the Sun and Earth as seen by the Voyager 1 spacecraft from a distance of about 6.4 billion km. The Sun appears like any other star, and its light washes out the tiny specks in orbit around it. Earth (inset) is nothing more than a blue speck in Voyager's eye.

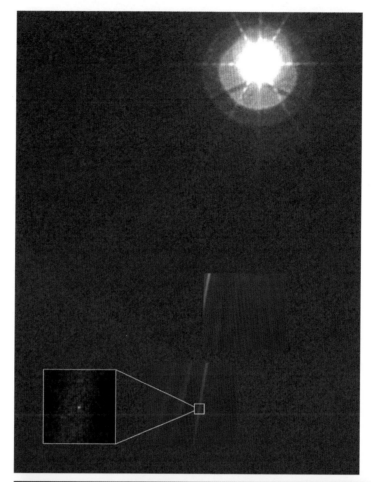

Figure 3.2. The Mauna Kea observatories – some of our most powerful eyes on the sky – are nearly invisible from low-Earth orbit. Can you spot the observatory domes at the center of the image?

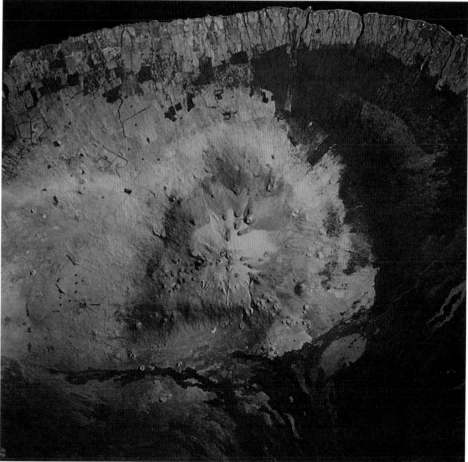

The changing solar system

The solar system is in a state of constant change. Every planet, moon, asteroid, comet, and even our Sun is subject to a variety of modifying processes. What are these forces that shape planets and keep legions of planetary observers busy? The most obvious – the easiest to spot in images – are impact cratering, volcanism, and atmospheric change. Evidence of these activities can be found in places like Earth's moon, the volcanic moon Io, the Great Red Spot on Jupiter, and the sculpted shapes of comet tails and planetary magnetospheres. In addition, studies of the motions of solar system bodies – their orbital dynamics – has, in recent decades, mated observations of smaller-than-planet-size bodies with more efficient mapping of the trips that these objects make throughout the solar system. In turn, we've all become more aware of the threat to our planet from comets and asteroids that stray into our own planet's orbital path. The worldwide effort in 1994 to study the effects of Comet Shoemaker–Levy 9's impacts into the Jovian clouds was a wake-up call, alerting us to the potential for asteroid and comet collisions with Earth. The search for objects that just might hit Earth goes hand in hand with our growing understanding of objects left over from the earliest epochs of solar system formation.

The large-scale evolution of the solar system took place long before anything was alive on our planet. The story began some 4.5 billion years ago in a cloud of hydrogen gas mixed with a good supply of heavier elements ejected from the deaths of massive stars called supernovae. Shock waves slammed into the material, compressing and shaping a primordial nebula. Continual compression heated the cloud and it began to rotate. The spinning motion flattered out the nebula and gravitational attraction pulled clumps of material together. The larger agglomerations attracted more and more material to themselves. The central lump became the proto-Sun. Elsewhere in the nebula, dust grains stuck together to form planetesimals, which collided and merged with one another to build proto-planets.

At some point, the core of the proto-Sun got hot and dense enough for hydrogen atoms to slam together and fuse into helium – a classic nuclear reaction. The hot young object had become a star. Fierce flares marked its infancy. Jets of radiation poured out like cosmic beacons heralding the new arrival. Winds from the newly born Sun nearly destroyed the nearby remnants of its gas and dust cradle, leaving four rocky bodies that eventually became airless Mercury, atmosphere-rich Venus, a life-bearing Earth, and enigmatic Mars. The larger planets, with their higher masses and gravities, managed to hold on to some of the primordial hydrogen for their own atmospheres. They became Jupiter, Saturn, Uranus, and Neptune. In the most distant, frigid reaches of the solar system, the coldest pieces of the primordial nebula survived the hot breath of the newborn Sun to become the icy worlds we know today as the Kuiper Belt – including the largest member, Pluto and its moon Charon.

Although the newborn planets swept up much of the material left in the proto-solar nebula, plenty of leftovers remained. Today these smaller bodies populate the solar system as comets and asteroids, and the many moons that orbit the planets. It has only recently been recognized that much of the flotsam is a storehouse of remnants from the early formation of the solar system.

This modern schematic of solar system formation is far from perfect. It leaves many unanswered questions about the origin of the primordial cloud and how long it took for the solar system objects to form. The theory does not completely explain why there's an asteroid belt between Mars and Jupiter, although we suspect that Jupiter's enormous gravitational influence prevented a planet from forming at the point where the asteroid belt lies. Nor do we have more than a hint at the size and extent of the reservoir of icy bodies that orbit at the outer limits of the solar system. What the theory does do for us, however, is give us an inside look at the processes of star and planet formation. That's an important first step in interpreting many of the objects we see in the sky.

Collisions and other catastrophes

The planets are worlds that exist because smaller bodies stuck together to make larger ones. Another way to think of this is that we are here because of impacts, and bodies smacking into other bodies is a way of life for the solar system. Living on a world with a blanket of air that gobbles up small, incoming rocks, we tend to forget that collisions still take place. In 1993, the world woke up to the news that Jupiter was about to be bombarded by Comet Shoemaker-Levy 9 in the summer of 1994. The idea of impacts isn't a new one, and the evidence of collisions is scattered across the solar system. This one, however, was the first one we could watch with a worldwide array of telescopes, almost in "prime time."

An extended network of scientists used observatories around the world, plus a collection of space-based instruments, to capture images and data of the event in visible, ultraviolet, and infrared wavelengths. No one on Earth actually saw the pieces of Shoemaker-Levy 9 plunge into the Jovian cloud tops because all the

Figure 3.3. In 1992 Comet Shoemaker-Levy 9 made a close approach to Jupiter. Stresses from the planet's strong gravitational field tore the comet's nucleus into 21 pieces and altered their orbit. The comet's path took it on a collision course with the giant planet in July 1994.

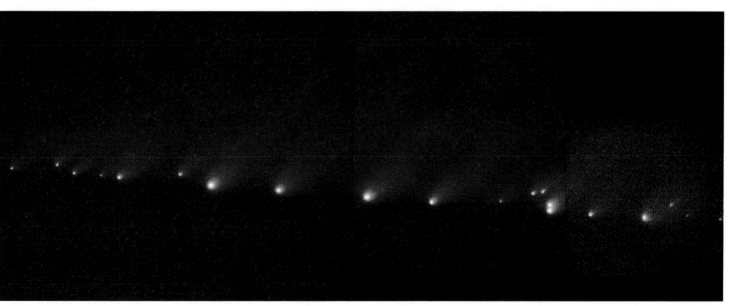

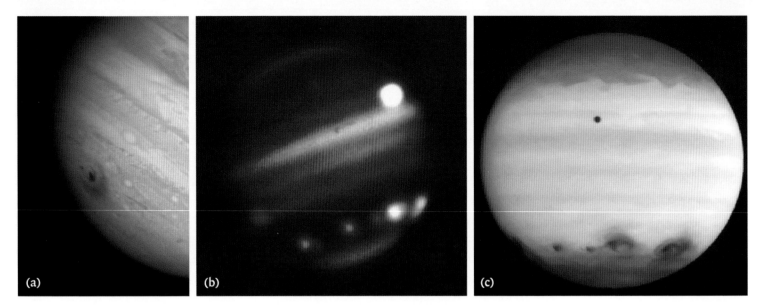

(a) (b) (c)

Figure 3.4. The successive impact sites on Jupiter were imaged using visible, infrared, and ultraviolet filters. **(a)** The large G fragment left a considerable scar that took days to dissipate. **(b)** Each impact generated hot spots, easily seen in the infrared. **(c)** An ultraviolet view shows the aftermath of several impacts. They appear dark because the impact kicked dust up into the atmosphere and obscured ultraviolet emissions.

action was occurring on the side of Jupiter facing away from Earth. But, due to Jupiter's rapid rotation rate, the impact sites came into view very shortly after the events occurred.

The Shoemaker-Levy 9 impacts raised public consciousness about the effects of such collisions, particularly on the heavily populated surface of our planet. In truth, Earth and every other body in the solar system has suffered impacts since formation. Some 4 billion years ago, about 1 billion years after the Sun first turned on, the inner worlds suffered a period called the Late Heavy Bombardment. It was the last large-scale rain of leftovers from the nebular cradle but this doesn't mean the solar system's birth debris is gone. Every year, the Sun vaporizes comets that chance to stray too close. Many of these chance encounters between the Sun and the frozen remnants of the proto-solar nebula are watched in real time by spacecraft such as the Solar and Heliospheric Observatory (SOHO) mission. The frequency of these death plunges confirms that the solar system is still quite heavily populated with comets and asteroids. The Sun's gravity draws them in like moths to a flame, sometimes from the outermost reaches of the solar system.

These death dives don't affect the Sun in any noticeable way, but if the same comets – or small asteroids – intersected Earth's orbit or some other solar-system object, the consequences would be very different. Over the course of geologic time, Earth's surface has been peppered with impacts, but the inexorable processes of weathering, erosion, and volcanism have masked many of the sites. Today there are at least 148 recognizable impact sites on Earth's surface. Some are eroded away or covered in vegetation, while several – like the 50 000-year-old Meteor Crater in northern Arizona – are quite obvious. A more recent event that occurred in 1908 in the Siberian region of Tunguska may have been the explosion of an asteroid before it could hit the ground. Photos taken of the area showed extensive damage to forests and witnesses reported seeing a fireball in the air just before the explosion took place.

The best-known impact site on our planet can't be seen – but its effect was devastating. It occurred some 65 million years ago, at a point in geologic history called the Cretaceous–Tertiary boundary. The result was the Chixculub Crater,

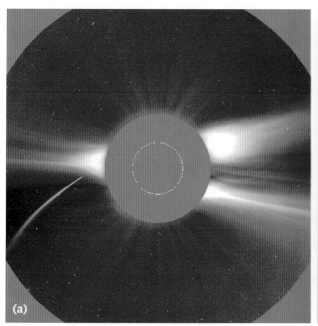 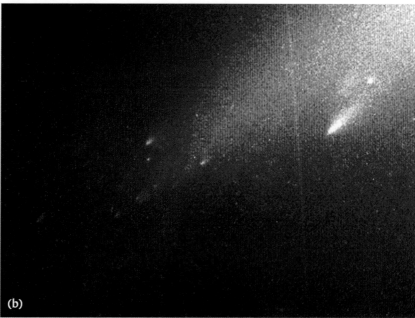

Figure 3.5. **(a)** The fate of a sun-grazing comet. The Solar Heliospheric Observatory's coronagraph caught this image of the comet SOHO-6 as it drew near the Sun on December 23, 1996. Here we see its dust tail streaming away from the nucleus. Eventually comet SOHO-6 was vaporized by the heat of the Sun. SOHO has discovered more than 500 comets during its mission. **(b)** Comets break up for a variety of reasons, spreading pieces of ice and debris along their former orbit. Hubble Space Telescope tracked Comet Linear (C/1999 S4) after it broke into pieces around July 26, 2000. HST's image was the first close-up look at cometesimals – the building blocks of cometary nuclei. The brightest fragment may have been the remains of the parent nucleus that fragmented into the cluster of smaller pieces.

buried deep beneath the Yucatan Peninsula and extending out under the Gulf of Mexico. The event that created Chixculub is widely considered part of a series of disasters that brought on the so-called "death of the dinosaurs." Similar impacts today could seriously disrupt life on Earth again. Combined with the graphic results of the Shoemaker-Levy 9 impacts on Jupiter, it is no small wonder that searches for near-Earth objects (NEOs) have taken on new urgency.

In the years since the Shoemaker-Levy 9 events, news of close approaches to Earth by various-sized asteroids have grabbed public attention. The effort behind the detection of these possible impactors is quite substantial – requiring a number of accurate observations in order to establish good orbits. This takes time, and the process is complicated by the fact that some of these bodies are too dim to be seen until they are quite close to us. A number of groups have formed to monitor the skies for possible near-Earth objects. Among them are an alphabet soup of groups and agencies, including the Asiago DLR Asteroid Survey (ADAS) and the Spaceguard Foundation (both in Italy), the Catalina Sky Survey at the University of Arizona, the Japanese Spaceguard Association, the Lincoln Near-Earth Asteroid Research project (LINEAR), the Lowell Observatory Near-Earth Object Search (LONEOS), project NEAT (the Near-Earth Asteroid Tracking mission, administered by NASA's Jet Propulsion Laboratory), the Spacewatch Project at the University of Arizona's Lunar and Planetary Laboratory, and others. As of this writing, NASA has a congressional mandate to discover 90 percent of all comets and asteroids that are larger than 1 km in diameter and have Earth-crossing orbits.

Impacts, collisions, and cratering are occupational hazards associated with living in a dynamic solar system. However, strange as it may seem, astronomers *can* help protect Earth by continually monitoring the sky for objects that might go bump in the night. Such continual vigilance can help identify threats and give us early warning of impending impacts. Here's how it works. When a new asteroid or comet is discovered in near-Earth space, observers in both the amateur and professional communities rush to make measurements of the object's path.

Figure 3.6. Earth's most famous crater as seen by the astronauts aboard the space shuttle. The Barringer Meteor Crater in northern Arizona formed when a piece of space debris about 60 m wide slammed into the Earth nearly 50 000 years ago. The impact excavated out 175 million tons of rock, and formed a crater a mile wide and nearly 600 feet deep. If this is what the collision of a relative small piece of space debris can do, imagine the damage if a larger asteroid had been involved! This crater is well preserved because its desert location greatly slows large-scale erosion from rain and floods. Craters in wetter climates would soon be erased, and impacts into the ocean would be largely undetected (unless they were large enough to cause tsunamis).

Predictions for where the object will be months or decades in the future can be uncertain until a large number of position measurements are taken and analyzed. Fortunately, for most objects, it quickly becomes clear that they are going to miss us. There are always a few for which accurate positions can't be determined, however, and if preliminary observations show that a given object might have even a slight chance of hitting us, astronomers assess the threat and warn the public.

The next step after discovery is "mitigation" – which in layman's terms means figuring out what steps to take to defuse the threat. Of course Hollywood-style fantasies of sending shuttle crews out with bombs to deflect asteroids immediately come to mind, but the more realistic approach would be less dramatic and more scientifically accurate. If astronomers can determine an object's orbit quickly and accurately, then theoretically a small probe equipped with thrusters could be sent out to intercept the object and gently nudge it into a slightly different Earth-*avoiding* orbit long before it got close enough to be a definite threat.

Figure 3.7. **(Opposite)** Cratering in the inner solar system involves mostly rocky bodies. A Clementine image of the cratered lunar surface **(a)** shows large impact basins peppered over with smaller craters. Mercury's ancient, pockmarked surface was mapped by the Mariner 10 mission **(b)**. Venus suffers from impacts as well. Although its atmosphere snuffs out the smaller impactors, larger meteorites dig out craters like the 69-km-wide Dickinson crater **(c)**. It is surrounded by material scooped out and melted during the impact. This picture is a radar image produced from data taken by the Magellan spacecraft. **(d)** Craters on Mars appear in all shapes and sizes. Galle crater (center) is often called the "Happy Face." Smaller craters and ejecta from the impact that scooped it out surround it.

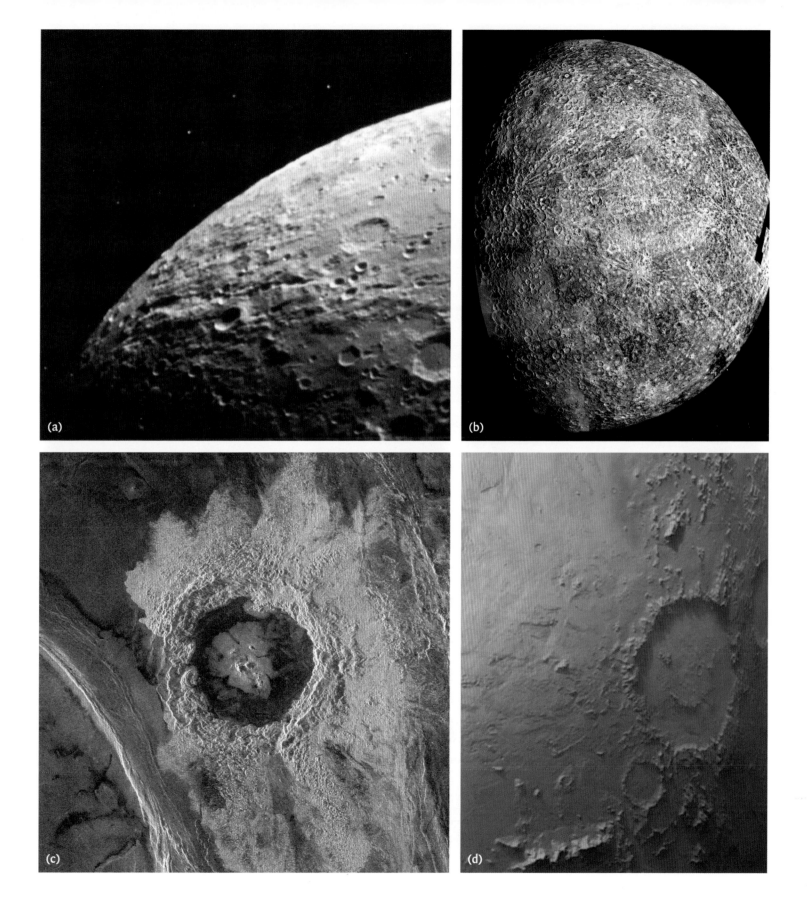

57

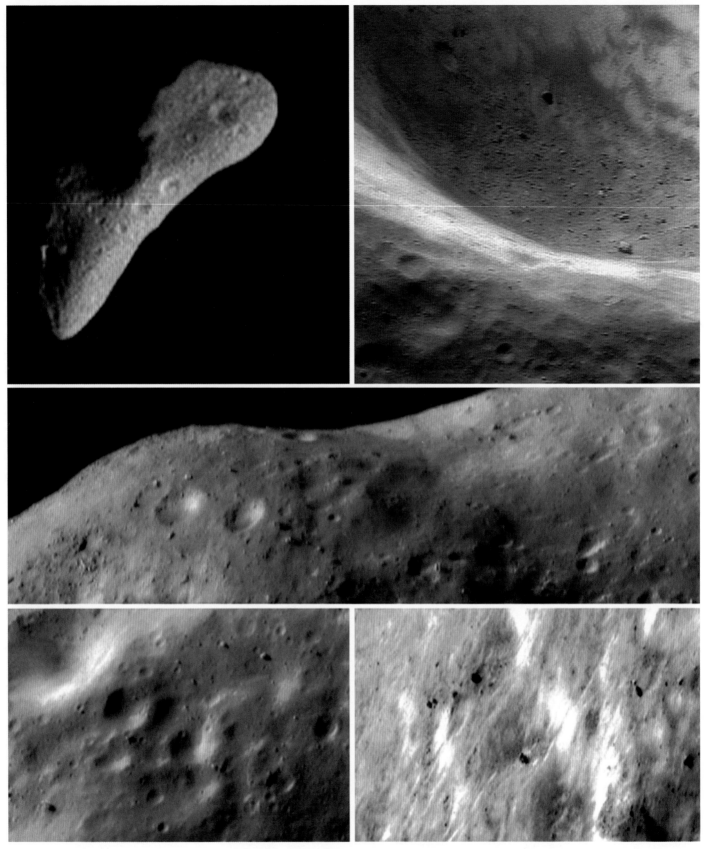

Figure 3.8. The Near-Earth Asteroid Rendezvous Shoemaker mission trained its cameras on the asteroid Eros **(upper left)**. Boulders are strewn around the floor of one crater **(upper right)** and over the surface **(middle, lower left and right)**. Its surface is littered with craters and a dusty powdery material called regolith. All of these views are false-colored to characterize the age and freshness of surface materials.

One outcome of this increased interest in near-Earth objects is a prediction scale that lets astronomers rate how threatening a given object might be. It's called the Torino scale and it rates an object's threat probability on a scale of 0 to 10, where 0 indicates an object has a zero or very small chance of colliding with Earth. (Zero is also used to rate any object that is not big enough to survive a passage through Earth's atmosphere, if it should get that far.) Obviously a 10 rating indicates that a collision is certain, and the incoming object will bring with it a global disaster upon impact.

When we look around at the other worlds of the solar system, it's obvious that impact events have occurred throughout the history of the planets. We find craters nearly everywhere – at least on the solid surfaces. Impacts into the gas giant planets never show up, or get erased fairly quickly by atmospheric change, as they did in the case of the impacts on Jupiter. Right next door to us, however, the Moon has a massive collection of impact sites. All you have to do to see them is step outside with a pair of binoculars and look at the lunar surface. Understanding of the impact process came about largely because of studies by scientists trying to decide if the lunar basins were volcanic craters or caused by collisions. As it turns out, the answer is "a little bit of both."

When an object slams into the surface of a planet or moon, a tremendous amount of energy is transferred. Due to the huge orbital velocities involved, a bullet-sized object would impact with roughly a thousand times the energy it would have when fired by a high-power rifle. The force of the collision vaporizes and melts the impactor as well as surrounding surface areas, and digs out the familiar crater "hole." The impact sends a rain of material called "ejecta" flying away from the site. Usually the ejecta are sprayed out onto the surrounding

Figure 3.9. Most of the icy satellites of the outer planets are pockmarked with craters. **(a)** Jupiter's icy moon Callisto boasts an ancient, crater-covered surface. **(b)** Saturn's ice world Enceladus has heavily cratered older terrain next to newer, resurfaced landscapes.

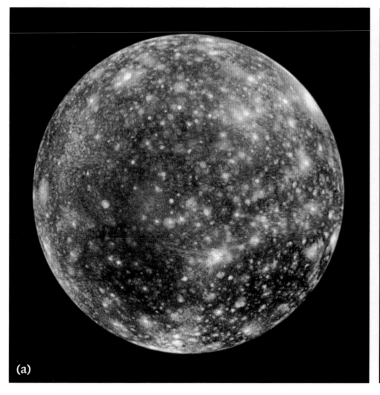
(a)

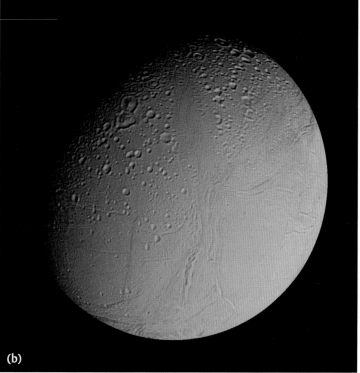
(b)

surface, although in some cases they splatter out to space, too. The melted material flows around the site – and in the case of the lunar impact basins (called "maria") – molten lava from under the surface welled up where the impactor punched through the surface. The resulting basins surrounded by ring-like mountains were first formed by an impact and then filled in with molten rock. On the icy moons of the outer solar system, the "lava" oozing up after impacts is more likely to be softened ice or even a slushy water–ice mixture. Subsequent impacts splash out new craters on top of the old. Throughout the solar system, some of the oldest surfaces we see are the ones with the most impacts, while others have been paved over by volcanism or the actions of wind, weather, or water.

Visitors from other worlds

In the range of impact dangers, meteorites landing on Earth are hardly in the same category as the large impactors that dug out the largest craters. Yet, the small bits of solar system debris that make it to Earth's surface often have quite interesting stories of their own to tell. Every year thousands of space rocks land on Earth, gouging out little more than a small depression where they fall. Most aren't associated with any known planetary body, but a few rare specimens appear to have come from the planet Mars and even the asteroid Vesta.

At first it seems mind-boggling that rocks from other planets could somehow find their way to Earth. To understand how it happens we point to yet another effect of impacts in the solar system: as a transport mechanism! Several million years ago, a meteoroid slammed into the surface of the Red Planet. Martian rocks were excavated in the process, and while most undoubtedly fell back to the surface, a few were sent rocketing off into space. They spent millions of years in space before intersecting Earth's orbit and eventually falling to our planet's surface as meteorites, and today researchers think that Mars meteorites reach our planet at a rate of about one a month.

Suspected Mars rocks were discovered when researchers compared them to rocks from Earth, and also with data returned by the Mars Viking missions in the late 1970s. Trapped gases inside the visitors matched the data Viking gleaned about the Martian atmosphere. As if that wasn't exciting enough, the Martian meteorites came to even more public attention in 1996 when a team of scientists asserted that certain small structures found within the rocks were evidence of early microbial life on Mars. A firestorm of interest and scientific criticism greeted this announcement and the research community went to work trying to prove (or disprove) the seemingly fantastic claims. As of this writing, the finer points of the research are still being debated among planetary scientists and astrobiologists, but the interest it has generated in astrobiology (the discovery and study of life elsewhere in the solar system) is tremendous.

Martian meteorites aren't the only "alien" rocks that can be traced back to a known solar system object. The asteroid Vesta appears to be the source of a collection of rocks called the HED meteorites (which stands for Howardites, Eucrites, and Diogenites) – families of meteorites that are related to each other by

Figure 3.10. **(Opposite)** On any clear night an observer can usually see a few meteors flashing across the sky. These flashes record the final fiery moments of a bit of space rock or dust as it plows through Earth's atmosphere and burns up. A few larger pieces make it to the surface and are picked up as meteorites. Meteor showers occur when Earth's orbit crosses paths with the orbits of dozens of comets that have left solid debris along their paths. This image captured the 2001 Leonid meteor shower during more than an hour's observing from a dark-sky site in Australia. To the lower right, the meteor shower radiates from the sickle-shaped string of stars in Leo. Orion (center left) lies near the Milky Way.

 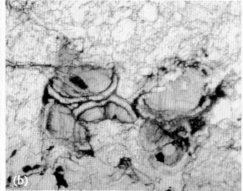 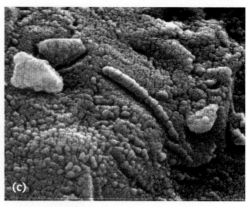

Figure 3.11. **(a)** A meteorite from Mars was found in the Allen Hills of Antarctica. Analysis of the rock told scientists that it had likely spent time underwater during Mars' earliest epochs. **(b)** A thin section of the rock showed banded carbonate mineral globules that imply liquid water existed on the Martian surface. **(c)** Further studies of the bands revealed microscopic shapes that some researchers think might be living and fossil bacteria, and the decayed remains and products of these life forms.

chemical composition. These rocky visitors formed from basaltic magma (a form of lava), but not on Earth. The implication is that their parent body must have experienced volcanic activity at some time in the past.

Vesta and its V-type asteroids illustrate the power of impacts and the orbital effects that make up the meteorite transport system between solar system bodies. From HST images and data, it's clear the asteroid suffered a major impact that blasted off a major chunk of crust and exposed deeper layers of material. Curiously, Vesta did not break apart during the impact, but major chunks of the crust were sent into orbit around the Sun.

How do we know that a given meteorite is from Vesta? Asteroids are classified into types based on their chemical spectra and surface brightnesses. Maps of Vesta show that much of its surface is made of igneous rock. This means that the upper layers of the asteroid were once melted, or possibly that lava flowing from its interior flooded the surface. Vesta is not a member of any of the more common types of asteroids, but its spectrum very closely matches the reflectance spectra of HED meteorites that astronomers think came from Vesta during the cratering event.

The orbits of the smaller V-type meteorite extend from Vesta (at 2.36 AU) to a point in space where an asteroid makes three revolutions around the Sun during Jupiter's year. This is known as a 3 : 1 resonance with Jupiter. Objects in such a situation quickly move into highly oblong orbits. Some are ejected from the solar system altogether. The major collision that produced the huge crater on Vesta knocked off multiple fragments that became the smaller V-type asteroids. After the impact, these pieces made their way out to the 3 : 1 resonance point and into highly eccentric orbits. It was only a matter of time before some of them intersected Earth's orbit, and wound up as meteorites.

The Sun and the polar glows

The Sun is the most important "shaper" of events and processes in the solar system. It is the most massive object in the system, keeping the planets, moons, and rings locked in their orbital paths. It's the main source of heat in the system, driving atmospheric change in all the planets. Its fiery breath – in the form of the solar wind – sweeps past at speeds of more than 400 km per second. It's a more or less constant windstorm, but it can turn blustery at times, buffeting the magnetic

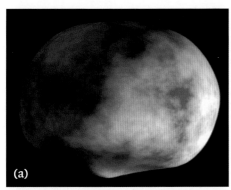
(a)

(b)

Figure 3.12. Vesta is the third-largest asteroid in the solar system. It measures 578 × 560 × 458 km and its orbit lies out past the orbit of Mars.
(a) This computer model of Vesta was synthesized from Hubble Space Telescope data. The asteroid is asymmetrical, probably due to an impact that carved out a 456-km-wide impact crater that gives the appearance that a big chunk is missing from the southern limb at the bottom of the picture.
(b) A piece of the asteroid Vesta was flung away during a collision, and eventually fell to Earth at Palo Blanco Creek, New Mexico.

Figure 3.13. An image of Jupiter's north and south polar aurorae, and a schematic representation of its magnetic field lines illustrates how a typical magnetic field emanates from a planet and guides the aurorae-producing particles. Most fields are generated deep within the core of the planet.

fields of the planets and causing kinky disruptions in comet tails. In addition, the surface of the Sun will often erupt in flares, emitting ultraviolet light and x rays.

All this frenetic activity can have a direct effect on the upper reaches of planetary atmospheres, resulting in displays of light called aurorae. Aurorae are caused by collisions between energetic electrons and atoms of gas in the upper atmosphere of a planet. The impacts "excite" the gases – that is, they cause the different atmospheric gases to glow in different colors. For example, excitation of oxygen causes greens, yellows, and deep reds to appear. Nitrogen produces blue and purple–red colors.

Our planet isn't the only place that sports belts of glowing gases around its polar regions. Aurorae have also been spotted at Jupiter and its moon Ganymede, Saturn, and Neptune. As on Earth, these aurorae are caused by the collisions of trapped electrons with atoms of atmospheric gases.

The discovery of aurorae on other planets and moons proves that those objects have magnetic fields. In fact, the most powerful characteristics of the gas giant Jupiter are not limited to just its size or its atmosphere, but its tremendously strong magnetic field. Trapped energetic particles – some of them from the moon Io – swirl around in this electrified environment, generating intense ultraviolet radiation. It is so lethal that spacecraft electronics can fry within seconds if they're not encased in a strong protective shield. These aurorae were first seen by the Voyager 1 spacecraft in 1979, and have been imaged by space and ground-based observatories ever since.

Changes in the brightness of the Jovian aurorae occur over the course of Jupiter's 10-hour day. One of the brightest features in the aurorae is something called the "Io footprint." As charged particles are caught in the Jovian magnetosphere and drawn to the poles in a flux tube, its "footprint" rotates across Jupiter at about 5 km per second as Io orbits the planet. If an observer could somehow be transported to Jupiter to see this phenomenon, it would be an experience unlike any auroral storm seen on Earth. At Jupiter's cloud tops, light would fill the sky, and the gases would glow at temperatures of more than 6000 degrees.

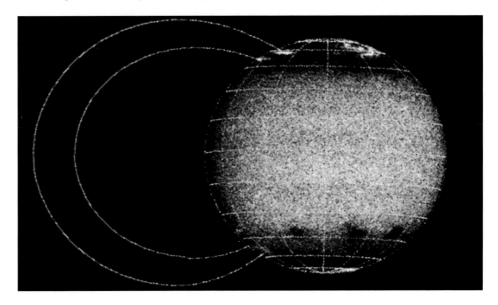

Figure 3.14. Earth auroral displays are regular sky sights for people living in the northern and southern regions of the planet – and on orbit too! Astronauts aboard the space shuttle Discovery in May 1991 captured this view of the southern aurora.

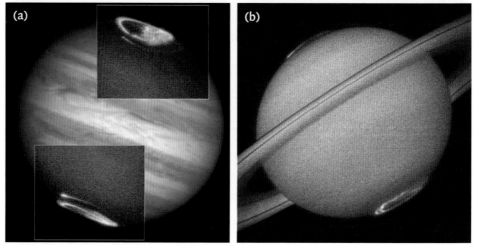

Figure 3.15. These false-color images are the result of Hubble Space Telescope observations taken with filters that permitted ultraviolet light to pass through. **(a)** Jovian aurorae can also form around Jupiter's magnetic poles when particles from the volcanic moon Io become trapped in the planet's magnetic field. Io's auroral "track" traces out a path equatorward of the regular auroral display at both poles. **(b)** Ripples, patterns, and brightenings in the Saturnian aurorae are all evidence of a powerful tug-of-war between Saturn's magnetic field and the solar wind.

More than two decades after the Voyager missions measured the aurorae at Saturn in 1980 and 1981, the Hubble Space Telescope, with its ultraviolet imaging capability, studied Saturn's aurorae. It also found evidence for auroral emissions at the poles of Jupiter's moon Ganymede. It is likely that underneath the icy surface of this moon there's a salty, electrically conducting ocean of water generating a magnetic field that interacts with Jupiter's more massive one.

COMPARATIVE PLANETOLOGY

Earth is the benchmark against which we compare all other planets. Perhaps we do this because it's where we originated and we are most familiar with its systems and processes. It stands to reason that if we see something we can explain on our planet, we should be able to explain its occurrence elsewhere. It's not just a question of figuring out how each planet is similar to Earth, but also how the worlds of the solar system differ from each other. One way to do this is to compare

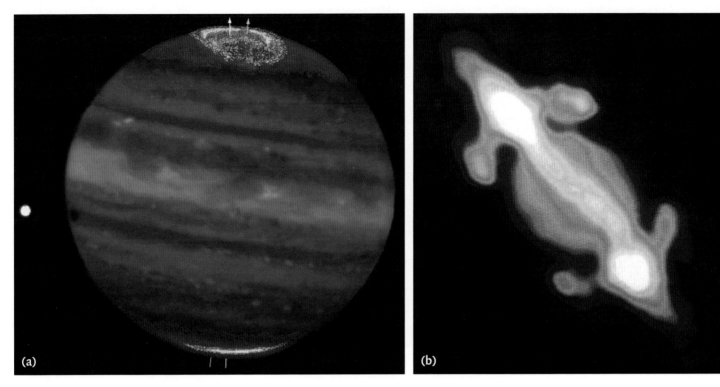

Figure 3.16. **(a)** A false-color infrared view of Jupiter taken from the European Southern Observatory in Chile shows the heat radiated by the Jupiter's aurorae. **(b)** A radio telescope sees Jupiter as an extended, almost tadpole-shaped object in the sky. Electrons trapped in the planet's magnetic field generate its amazingly strong radio signals. Jupiter lies in the center of the oblong central region. The outer "wings" are evidence of radio emissions by charged particles in the planet's magnetic field.

what structures and processes they have in common – such as the impact events we've just discussed.

The planet that seems most like Earth to us is Mars. Let's look at it as a simple exercise of comparative planetology. Its canyons and craters, polar caps, and ever-changing cloud wisps have always fascinated humans. The planet is close enough that faint surface markings can be made out through a small, backyard telescope. Over the course of a Martian year, observers can watch the planet's polar caps grow and shrink with the change of seasons. At certain times of year, storms nearly obliterate our view of the surface.

We have inherited some amazing cultural baggage about the Red Planet. In early religions, Mars was a god of war, something to be feared. At the middle of the twentieth century, before our first missions to the Red Planet, Mars mania focused not on belligerent deities waiting to mete out punishment on cowering humans, but on science-fictional adventures like *The War of the Worlds* or the adventures of Edgar Rice Burroughs's John Carter and the Martian princess Dejah Thoris. These adventure stories dealt with life on Mars.

As we all know, the exploratory flights of robotic probes put to rest some of our flights of fancy regarding the possibility of life on Mars. That's the way science sometimes works – it takes a perfectly nice little fantasy and injects a little reality. In this case however, the reality of Mars makes it an even more tantalizing place to visit – and for some would-be Mars explorers, trips to the fourth planet can't come fast enough!

To understand what happens on Mars it helps again to look at Earth as a benchmark. There are places on our planet that do a fair job of mimicking the Martian surface. Haughton Crater in Devon Island in the northern arctic regions is a good example. For the past several years, teams of scientists have lived and worked in this desolate, frozen desert. Explorers have also set up

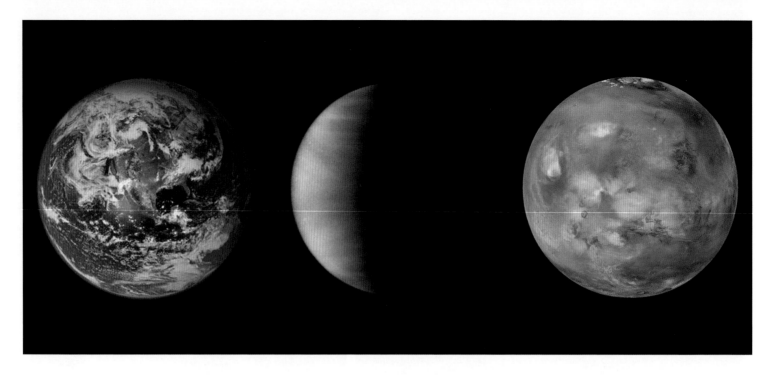

Figure 3.17. **(Left)** Watery, blue Earth as seen by the Clementine spacecraft. Mexico is centered in the field of view. Cloud-shrouded Venus **(center)** was imaged in ultraviolet light by the Hubble Space Telescope. Mars Global Surveyor's global map of Mars **(right)**, made from data taken during 12 daily orbits shows bluish-white water ice clouds hanging above the Martian volcanoes. A closer look reveals impact craters, a giant rift canyon, polar caps that grow and shrink with the change of seasons, and broad windswept plains. Noticeably lacking is any free-flowing surface water, although Mars researchers think that much of the Martian water supply may be locked in sub-surface ice called permafrost.

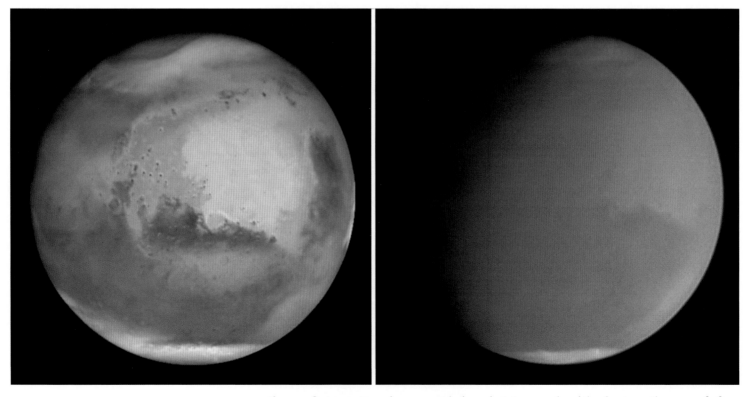

Figure 3.18. In 2001 Mars observers watched as a dust storm enveloped the planet over the course of a few days. The storm lasted for several months, hiding all but the most prominent features from view.

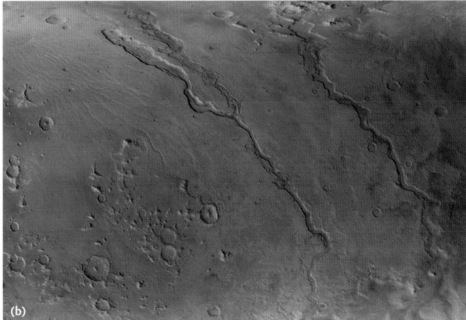

Figure 3.19. Comparative planetology gives us a chance to contrast features on Mars with features on Earth. Knowing how a landform was created on Earth helps us determine how it may have been formed on Mars. Try to pick out what appear to be flow channels in these two images. Only a telltale spot of green in the upper right corner of **(a)** gives away its planetary location – in this case, an irrigated field in the sand hills of Nebraska. **(b)** Three valleys on Mars: Dao Vallis, Niger Vallis, and Harmakhis Vallis. These river valleys may have formed in the distant past by catastrophic floods of liquid water.

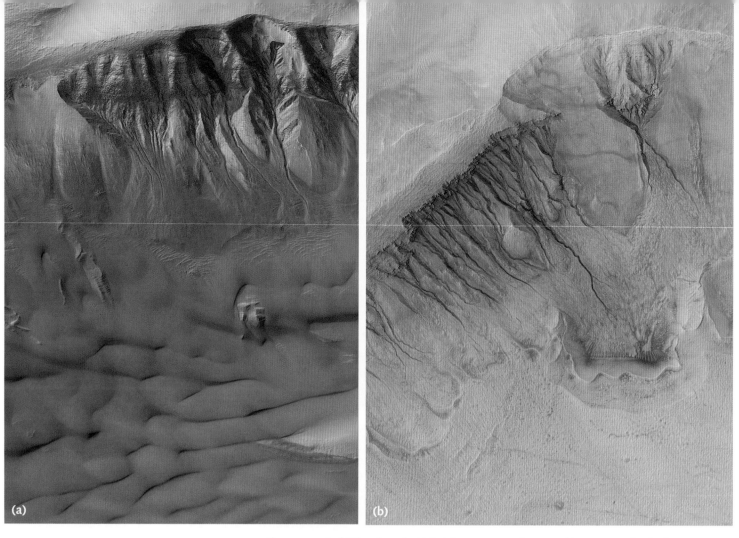

(a)

(b)

Figure 3.20. **(a, b)** There is no doubt that the movement of water on Mars at some time in the past contributed to the modern-day look of the planet. Detailed Mars Global Surveyor images of two different impact sites in Sirenum Terra show familiar-looking landforms notched into the crater rims. The gullies may have been formed by the release of groundwater over the Martian surface. Even though we don't know exactly when these events occurred, they look similar to water-eroded gullies on Earth. Study these images for a few minutes to find other familiar-looking landforms like sand dunes and debris fields littered with rocks that tumbled down the slopes to the floor of the crater.

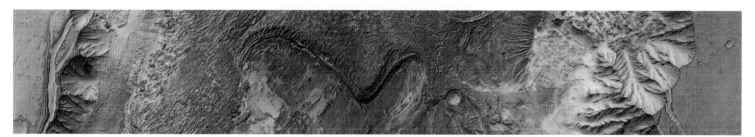

Figure 3.21. The Mars Odyssey spacecraft continually scans the surface with a Thermal Emission Imaging System (THEMIS), an instrument sensitive to wavelengths of light in both the visible and infrared parts of the electromagnetic spectrum. It is being used to study the mineral makeup of Mars dust and rocks. Color differences in this false-color infrared image represent differences in the composition of Mars surface materials. The image shows a portion of a canyon named Candor Chasma within the great Valles Marineris system of canyons. If liquid water existed here at any time in the past, it would have an effect on the mineralogy of the rocks. Any deposits of water ice on the surface would show up in the infrared as dark patches.

Figure 3.22. Ground truth at Ares Valles. It's one thing to study and map a planet from above, but the way to learn the most about a planetary surface is to land there and walk around. But for the fact that this scene lies some 70 million km away, this could be a desert landscape on Earth. In 1997, Mars Pathfinder bounced to a landing in this immense rock garden in Ares Valles. A long-gone flash flood tore across this plain, scattering rocks across the landscape. Windblown sand and dust lie in drifts between the rocks. In the distance are two low hills nicknamed the Twin Peaks. To date all the successful Mars landers have settled to ground in relatively safe flat areas. While these have provided a wealth of information and photogenic visions of the Red Planet, some truly fascinating places lie in the canyons of the Valles Marineris and atop the Martian volcanoes – places of considerably more danger to a fragile landing craft.

bases in Antarctica, and more such "Martian training camps" are planned for other places on Earth. To form the most accurate picture of changes on Mars, however, we need to monitor the planet itself over many years in as many ways as possible. That means visiting the planet, as well as observing it using HST and ground-based observatories. Our meticulous *in situ* explorations of Mars began in the 1970s and continue to this day. Even as this is written two spacecraft are orbiting the planet, continually mapping and photographing the dry, dusty surface, and more are on the way.

Mars is a world with a tormented past. The evidence is everywhere. Aside from the profusion of craters, canyons, plains, and broken terrain that mar the surface, many of the questions we have about the Red Planet are aimed at finding evidence for a warm, wet early Mars with a thicker atmosphere than it has now. If it *did* have more water (and a more substantial atmosphere) in its infancy, the next question to ask is whether it began with the same amount of water as the Earth did. If so, then what happened to that water?

There is no question that water once graced the surface of Mars. Lengthy flow channels with tributaries spread across the dusty plains, looking remarkably similar to dry riverbeds on Earth. Ancient, rock-strewn plains reveal a landscape tortured by the catastrophic rush of water across the surface. Where was the water flowing to? And where is it now? Scientists think that some small fraction of the original water supply is locked underground in permafrost, but it seems certain that a great deal of it escaped into space. Mars researchers are still debating how water could ever have existed on the surface given the low atmospheric pressure. It's not clear whether that water flowed across the surface in catastrophic floods, or seeped underground and undermined the surface.

Mars – with its thin atmosphere and absence of surface water – presents us with a challenge. How do we explain the process by which a world that apparently was once so wet lost its water and much of its atmosphere over a short period of time? Is it possible that it never had water on the surface to begin with? These are questions that planetary scientists are trying to answer as they pore over the wealth of images and data returned by the ongoing Mars mapping missions.

Volcanoes of the solar system

Volcanism is a good example of a world-shaping process that seems to be in operation throughout the solar system. We're all familiar with the volcanoes of Earth. They actively resurface not just the continents but also the sea floor. They

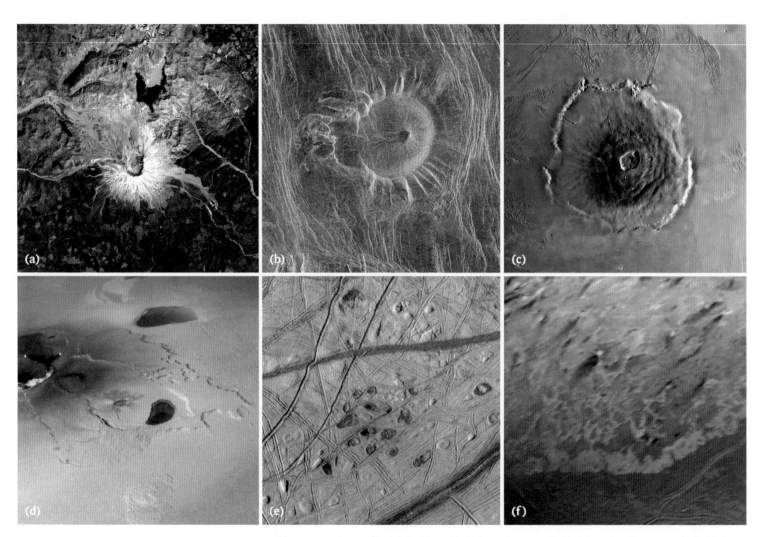

Figure 3.23. From orbit, **(a)** the Mount St. Helens volcano on Earth, **(b)** a volcanic structure in the Eistla Region on Venus, and **(c)** Olympus Mons on Mars represent the different forms that volcanic activity can take on these worlds. **(d)** On Io, a caldera in the Tvashtar Catena chain of volcanoes is spewing hot lava across the surface. The two small bright spots are sites where molten rock is exposed to the surface at the toes of lava flows. The larger orange and yellow ribbon is a 60-km-long lava flow that is cooling down. Io's continual volcanism spews clouds of sulfur out across its surface, and sends waves of heated particles into orbit around Jupiter. **(e)** The surface of Jupiter's icy moon Europa has been described as churning like a "lava lamp." Something beneath the surface of the moon is warming the icy crust, and sending warmer ice moving up from below. Frozen upper layers of colder ice sink down toward the moon's core. The reddish spots and shallow pits that deform the enigmatic ridged surface of Europa in this view may hold clues to the composition of Europa's melted ice ocean. **(f)** Out at Neptune, icy Triton sports volcanic activity in the form of geysers with dark plumes that look like black smudges across the surface.

bring molten rock up from deep beneath the Earth's crust, and are responsible for belching up some of the gases that make up our atmosphere.

It turns out that volcanoes disrupt surfaces and add gases to atmospheres throughout the solar system. Volcanism occurs when a world is heated from within. On Earth and its two sister planets Venus and Mars, the heating is caused by the decay of radioactive elements deep within the core. This is called radiogenic heating. There is also additional heat left over from the formation of each planet. On the inner planets, the heating forces molten rock (called "magma") out onto the surface, along with a variety of gases that mix into the atmosphere. The lava flows can be very fluid, or very rough and blocky. Sometimes, as in the case of the infamous explosions at Mount St. Helens in 1980, a volcano will blow itself apart, sending a rolling cloud of molten rock, ash, and gases over the landscape. Because Earth's crust is essentially a series of plates riding along over the mantle layer (which itself encloses the molten core), we can often find volcanoes at the boundary between two plates. In other areas, hot spots in the crust melt just enough to allow magma to bubble through to the surface. The Hawaiian island chain has formed as the Pacific plate rides over just such a hot spot.

Until the Magellan mission to Venus in the 1990s, no one quite understood the extent of volcanism on that cloud-shrouded world. While the inner core of the planet is very similar to Earth's, it has only one crustal plate underlying the surface, instead of several as Earth does. The volcanic flows on Venus are made up of very fluid lava that keeps the surface hot and pliable. In some places, the flows pile out into mountains and some very unusual pancake-shaped forms.

Volcanism at Mars suggests that the Red Planet was once a very active world and that it may not be dead yet. Like Venus, it does not appear to have plates overriding its mantle and core. Instead, a series of volcanoes are built up over weak spots in the crust. The tallest of these is Olympus Mons, which rises up some 27 km into the thin atmosphere. There don't appear to be any ongoing Martian eruptions now, but evidence of prior Martian volcanism points to an active past.

When we move out beyond the orbit of Mars, the face of volcanism changes abruptly because instead of studying planets, we're focused on several of the moons orbiting around gas giants. The most famous of the small volcanic worlds is Jupiter's rocky moon Io. Before the Voyager 1 mission uncovered proof of its fiery nature, scientists suspected that volcanism was occurring there, but the extent of activity that unfolded in the Voyager data was stunning. The two Voyager flybys resulted in images of at least eight eruptions and six volcanoes.

Io is squeezed into an orbital resonance between Jupiter, Ganymede, and Callisto. This puts it under exceptionally strong gravitational influences and causes the whole moon to flex. That, in turn, heats the interior and ultimately melts the rock. The "melt" reaches the surface in the form of sulfur flows, and fountains of sulfur dioxide erupt into the tenuous atmosphere.

When the worlds turn icy – as with the Jovian moon Europa or Neptune's moon Triton – volcanism becomes a process by which ice is transported to the surface. At Europa warmer ice makes its way up through a slushy ocean and disrupts the surface. Colder ice then slowly submerges itself back into the layer of slush. Researchers aren't quite sure what heating mechanism is causing this blue little world to stay relatively warm and resurface itself with ice volcanoes while other bodies around it are more solidly frozen. It may be the same tidal interactions that keep Io in a plastic state, or possibly radiogenic heating from a small rocky core. At Triton, the Sun warms the methane ice crust, which in turn heats the underlying nitrogen ice layer. Small volcanic vents act as pressure valves that allow nitrogen gas and a mixture of organic compounds to be ejected into the thin atmosphere.

My atmosphere's bigger than yours

Astronomers are fond of grouping objects into categories, particularly in the solar system. There are many different bins you can sort solar system objects into – and which ones you pick depends on what you want to study. We have the so-called "terrestrial planets" in the inner solar system: Mercury, Venus, Earth, and Mars. Then there are the gas giants Jupiter, Saturn, Uranus, and Neptune. The icy worlds comprise Pluto, Charon, the Kuiper Belt objects, and the larger satellites of the gas giants. The smaller moons, rings, asteroids, comets, and meteorites occupy their own bins in the cosmic sorting process.

Another way to sort worlds is to divide them up into those with magnetic fields and those without. Or, if you're looking for visible characteristics, sorting into "places with atmospheres" and those without any gassy envelopes is very useful. This puts Earth, Venus, Mars, Jupiter, Saturn, Uranus, Neptune, Io, Europa, Titan, and Triton into one bin, and everything else into another. But this leaves out Pluto, which grows an atmosphere during parts of its 248-year-long "year," and comets, which sprout clouds of gases as they get closer to the Sun and lose them as they move away. These are possibly the most extreme examples of atmospheric change we have in the solar system. By comparison, everything else is a study in similar structures yielding amazing contrasts.

As you examine the planets this way, one thing becomes immediately obvious: Earth's atmosphere is not exactly like the others. Why then, do scientists insist on using our planet as the benchmark against which we measure all the other planets? It's true that we get our concepts of the layered structure of atmospheres from a study of our gassy envelope. Even the thinnest ones we see at places like Mars or Triton are thicker at the surface and thinner at the top. But we are not so odd. Four planets are surrounded by what amount to massive deposits of hydrogen and helium in various forms. One planet – Earth – has mostly nitrogen and oxygen, Venus is mostly carbon dioxide with passing wisps of sulfur dioxide, and Mars is also a carbon dioxide atmosphere with traces of oxygen and water vapor. If you had to group by similarity and define the atmospheres of the solar system that way, the "typical" atmosphere would be like those at Jupiter, Saturn, Uranus, and Neptune.

Jupiter is a world of superlatives: strongest gravity, largest atmospheric storms, strongest magnetic field, most colorful clouds, most massive atmosphere, largest moons, numerous smaller moons, and thinnest ring system. Dig deep into its structure and below those clouds exist layers of molecular hydrogen and liquid metallic hydrogen. Below them lies a small solid core.

Figure 3.24. (a) The Cassini–Huygens spacecraft cameras returned this Voyager-quality image of Jupiter as it approached the planet. (b) The Galileo probe produced a highly detailed close-up the Great Red Spot, a Jovian hurricane that whirls its way through the atmosphere of the planet.

Figure 3.25. In 1995 a probe from the Galileo spacecraft entered the upper cloud decks of Jupiter. As it dropped deeper into a "hot spot" near the equator of the planet, the probe transmitted information about the region's chemical makeup, temperature, and atmospheric pressure. This view is a computerized three-dimensional visualization of the data returned by the probe. We are looking northeast across 34 000 km, from a viewpoint between two cloud layers. The warm region is clearly visible as a deep blue feature. Clouds traveling along in streak lines descend and evaporate as they approach the hot spot and the upper haze layer is being pushed up slightly by the heat pressure.

However, Jupiter's most striking feature is its atmosphere. Belts and zones seem to divide the planet's appearance into colorful cloud strata. Storms of all shapes and sizes reel across the top of the atmosphere as the clouds blow past each other. This constant seething motion produces incredible wind shears and displacements in the cloud zones of the planet. These displacements were of great interest to the Galileo science team members as they prepared a special probe to drop into the Jovian atmosphere. During the more than hour-long descent, the probe measured temperatures, pressures, density, gas composition, cloud layers, energy flux, radiation, and lightning in the first 1200 km of the upper cloud decks.

The data from the Galileo probe changed some very long-held views of the Jovian atmosphere. Analysis of the data found that Jupiter's winds, which blow more than 500 km per hour, stretch deeply into the thick atmosphere, and are not limited to the upper levels of the atmosphere. Scientists were hoping that the probe would measure the temperatures and densities of the three layers they expected to find, but the probe didn't find any evidence of such clearly defined regions. It's possible that they do exist, but that the probe's entry point was at an area drier and warmer than expected.

Figure 3.26. Saturn appears distant and serene in the first color composite made of images taken by NASA's Cassini–Huygens spacecraft as it approached the planet in October 2002. Titan, Saturn's largest moon, appears in the upper left. Titan is a major attraction for scientists of the Cassini–Huygens mission. The spacecraft will release the piggybacked Huygens probe, which will descend through Titan's atmosphere in early 2005 and report on conditions at the surface.

Jupiter's most obvious atmospheric feature is the Great Red Spot. It stares out from the planet like an angry eye. It's a huge high-pressure storm – a Jovian hurricane visible from Earth – and has whirled around in the southern hemisphere of the planet for at least 400 years. It's large enough that three Earths could comfortably fit across its width.

The gas giant Saturn is best known for its glittering ring system and has long been a favorite of both amateur and professional astronomers. Along with the other gas giants, however, it sports a massive atmosphere of hydrogen overlying oceans of liquid metallic hydrogen, all supported by a small rocky core. Saturn's atmosphere is likely as active as Jupiter's, but a haze layer tones down the look of the cloud tops to muted yellows and browns. Given our interest in the planet, it should come as no surprise that Saturn was selected as one of the first planetary objects to be studied by the Hubble Space Telescope. In fact, Saturn presented HST with its first solar system challenge only 4 months after launch.

On August 26, 1990, HST took what many called a "Voyager quality" image of the planet. It showed Saturn as it would have appear to us if it were only twice as far away as the Moon. In reality, Saturn is 1.4 billion km away – or about 3700 times farther away than the Moon. HST's image captured the northern hemisphere of Saturn, showing the banded structure in the planet's upper cloud decks. The famous Cassini Division (the largest dark gap) in the rings was clearly

Planets on a pixel

(a)

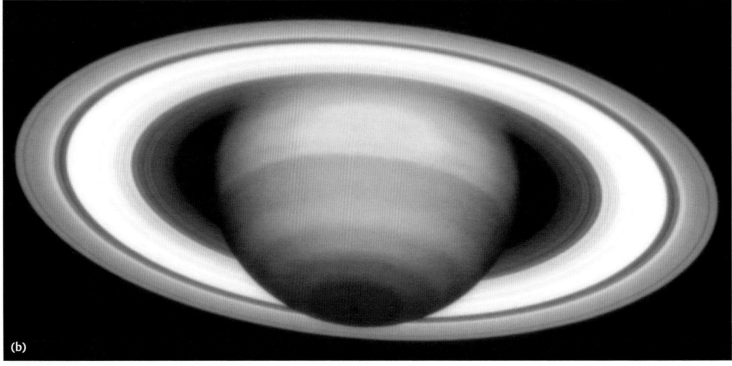

(b)

76

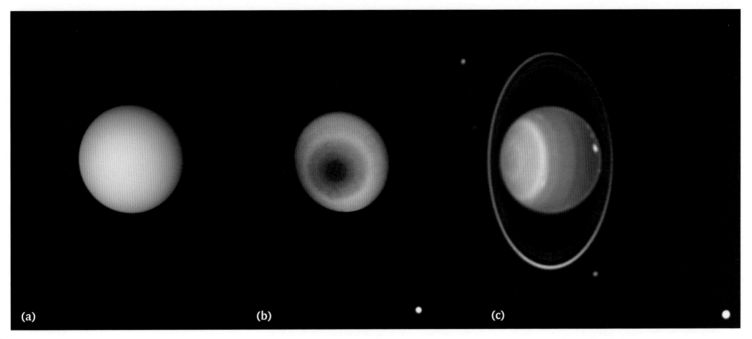

(a)   (b)   (c)

Figure 3.28. The Voyager 2 mission rounded out its gas-giant flybys with visits to Uranus in 1985 and Neptune in 1989. Hubble Space Telescope has zeroed in on the two outermost gas giants several times, charting changes in their atmospheres and studying both planets with its infrared-sensitive instruments. **(a)** In visible light, Uranus is little more than a featureless blue ball, but when Voyager's cameras used special filters to peek beneath the blue–green methane haze in the upper atmosphere astronomers were able to construct a false-color image **(b)** highlighting structure in the lower clouds. The dark polar hood and lighter bands are atmospheric phenomena similar to smog caused by chemical reactions set off by ultraviolet light from the Sun and atoms of gas. **(c)** A Hubble Space Telescope infrared view of Uranus highlights storms and the banded structure of the upper atmosphere and a thin set of rings. The clouds (on the right limb of the planet) circulate at speeds of up to 500 km per hour high in the planet's stratosphere.

Figure 3.27. **(Opposite) (a)** This image of Saturn from Hubble Space Telescope shows a rare equatorial storm raging in the upper cloud decks of Saturn's atmosphere. The arrowhead-shaped storm was generated by an upwelling of warm gas from the lower cloud layers of the planet. **(b)** Several years after Hubble Space Telescope took its first images of Saturn, the ground-based Very Large Telescope in Chile captured HST-quality images of Saturn using adaptive optics and special infrared filters to sharpen the view. This is one of the sharpest ring system views ever achieved from a ground-based observatory. The main ring sections are all easily visible, plus the Cassini Division (the dark band that seems to split the rings), and the Encke Division (a smaller gap just inside the outermost edge of the rings).

visible, as was a thin gap near the outer edge of the rings called the Encke Division. Before that time, the Encke gap had never been photographed from the Earth, making HST's accomplishment a rare one indeed.

About a month after HST's first image of Saturn, amateur observers announced the discovery of a large white spot developing in Saturn's northern equatorial region. Immediately, the observing community clamored for more time on the telescope to observe this massive atmospheric occurrence. After a hurried few weeks of planning, HST peered at Saturn, and took an image. It was an exciting time for the planetary science community whose last detailed look at Saturn was during the 1981 Voyager 2 flyby mission. Since then, HST has been the major space-based instrument of choice for observing the ongoing changes in Saturn's atmosphere. That's about to change. The Cassini–Huygens spacecraft, on a 4-year mission to thoroughly explore the ringed planet, will enter Saturn orbit in July 2004 and begin studying the planet and its moons and rings.

Our first enduring images of Uranus and Neptune as gas giant worlds in their own right came from the flyby images taken by Voyager 2 in 1986 and 1989. Their structures are similar to Jupiter and Saturn, with heavy atmospheres blending down into layers of liquid hydrogen and helium. Below that strange mixture is almost certainly a layer of water (probably in some form of ice), and beneath that a

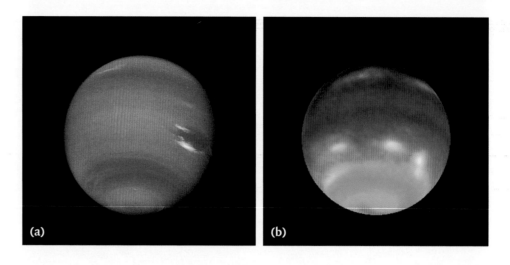

Figure 3.29. (a) Neptune also sported a methane haze when Voyager imaged it, and showed off a huge storm reminiscent of Jupiter's Great Red Spot. The Great Dark Spot, as it came to be called, was nowhere to be found a few years later when Hubble Space Telescope looked at the planet. This disappearing storm raises questions about Neptune's atmospheric conditions. **(b)** Combined Hubble Space Telescope and ground-based infrared images of Neptune show banded structures in the upper atmosphere of this gas giant planet.

(a)

(b)

rocky core about the size of Earth. Eight years after Voyager 2's farewell mission to the outer planets, astronomers began a program of Uranus observations with HST. Because the planet is tipped on its axis about 90 degrees with respect to the solar system, Uranus literally rolls around on its side as it orbits the Sun. During parts of its orbit, this causes the poles to be heated, rather than the equatorial regions as with the other planets. The ongoing observation program has focused on detecting seasonal changes in the high-altitude hazes that form at the poles and blanket the planet. Hubble recently found about 20 new cloud features at Uranus, indicating that atmospheric change is alive and well there.

During the Voyager 2 flyby Neptune distinguished itself by sporting a set of storms. These storms raced across the planet's disk, at wind speeds as high as 325 m per second. Along with the storms, a series of high-altitude, cirrus-like clouds set Neptune apart from its distant neighbor.

Recent studies of Neptune have combined simultaneous observations made with the HST and NASA's Infrared Telescope Facility in Hawaii. Neptune's weather continues to bluster around, creating new storms and boosting winds at the equator up to 1450 km per hour. No one is quite sure what drives these near-supersonic winds and creates giant storms that rage across its surface. It may well be some interaction with the dim, distant Sun and some processes going on deep in Neptune's layers. Whatever it is, it causes storms to form and disappear, rather than live on for decades or centuries as they do at Jupiter. However, Neptune does have one structural feature in common with the larger gas giants – a series of weather bands that run parallel to its equator.

Little worlds and big rings

Planetary moons and ring systems have come into their own as fascinating places to study. In the grand scheme of solar system formation, the smaller worlds, for example, occupy an interesting evolutionary niche. Earth's satellite was likely formed as the result of an impact between the newly formed Earth and a Mars-sized impactor. The resulting chaos resulted in the agglomeration that became our lunar neighbor. Other moons, like Phobos and Deimos at Mars, are captured asteroids. In the outer solar system, many of the largest icy moons that swarm around the gas

Figure 3.30. The moons of Mars – **(a)** Deimos and **(b)** Phobos – are small, oddly shaped, and cratered. It's quite likely they are captured asteroids.

Figure 3.31. A multiplicity of moons circles Jupiter. The largest – **(a)** Ganymede, **(b)** Io, **(c)** Europa, and **(d)** Callisto – were studied at length by the Galileo spacecraft. The fine cracks that seem to split the surface of Europa in this image are probably caused by stress on its icy crust. Slushy water ice from underneath wells up in the cracks and freezes onto the surface. Callisto's darkened ice surface is pockmarked with fresh ice splashed out as objects collided with it. It may harbor a salty, electrically conducting ocean under that pockmarked surface. Ganymede appears to have been hit recently, and the older areas on its surface show the remnant of a large impact event sometime in the distant past. Io is a continually venting volcanic world in the act of resurfacing itself.

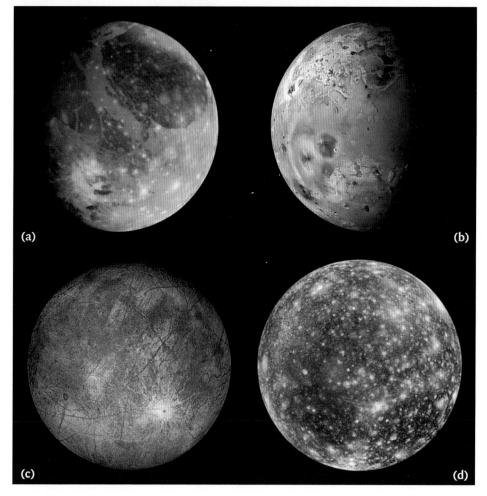

giant planets most likely formed when the planets did. Others may be parts of larger worlds that were destroyed by strong gravitational influences, or captured cometary nuclei on their way in from the outer realms of the solar system.

As discussed earlier, both Io and Europa are under intense scrutiny – both because of their unique forms of volcanism. Astronomers using HST discovered

Figure 3.32. Voyager images of Saturn's moons revealed dozens of satellites in orbit around the planet. The largest are shown here. From left to right, top to bottom they are: Titan, Rhea, Iapetus, Enceladus, Tethys, Mimas (with its death star-like crater), and Dione. Titan shows a few details in the clouds that hide its surface. The southern hemisphere appears lighter in contrast, and a well-defined band is seen near the equator. A dark collar seems to circle the north pole. All these bands are associated with cloud circulation in Titan's atmosphere. A hazy layer similar to the smog that sometimes blankets parts of Earth covers the clouds. The icy moons show evidence of cratering and resurfacing, leading scientists to wonder about the mechanisms at play beneath their surfaces.

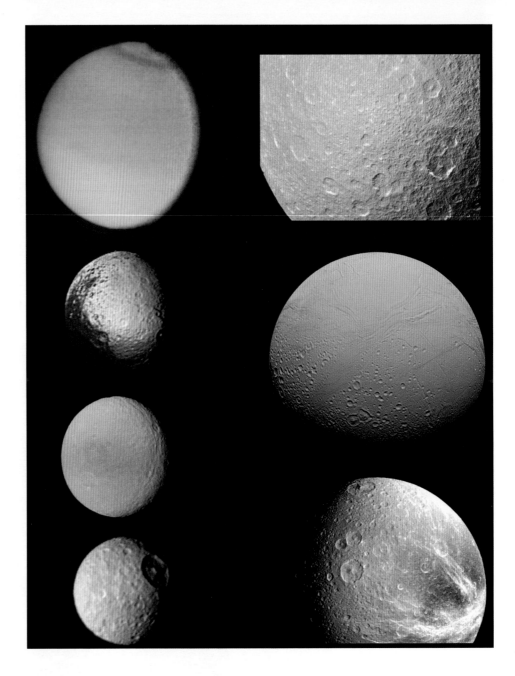

that Europa has a tenuous atmosphere of molecular oxygen. On Earth, the presence of oxygen indicates the presence of life, but on Europa, the discovery of oxygen points to some non-biological process – possibly related to impacts from dust and charged particles onto the water ice surface. Heating from these processes is just enough to cause the frozen water ice to melt, producing water vapor and molecules. A series of chemical reactions ultimately produces hydrogen (which escapes to space) and oxygen, which manages to accumulate and form an atmosphere extending about 200 km above the surface. Coupled with in situ observations of Europa's cracked, icy surface by the Galileo spacecraft, scientists will use HST observations to understand the interior structure of this world – and answer questions about whether it could harbor and support life.

The Galileo spacecraft has mapped the major surface features on Callisto and Ganymede. Impact craters and light and dark areas (called albedo features) of

Figure 3.33. The moons of Uranus and Neptune are icy worlds that show evidence of cratering, resurfacing, and geyser action. **(a)** The Uranian satellite Miranda exhibits a variety of surface units, indicating that this moon may have been broken apart and reassembled. **(b)** Ariel has huge cracks running across its surface, and a number of impact craters. **(c)** Titania, **(d)** Oberon, and **(e)** Umbriel (upper right) are also cratered, and dark Umbriel sports a ring of what appears to be fresh ice around its pole. **(f)** Neptune's largest moon, Triton, also has a variety of surface units, including the so-called "cantaloupe terrain" that stretches across the diameter of the moon and appears faintly blue in this image. The dark streaks are geysers shooting plumes of material out from just beneath the crust.

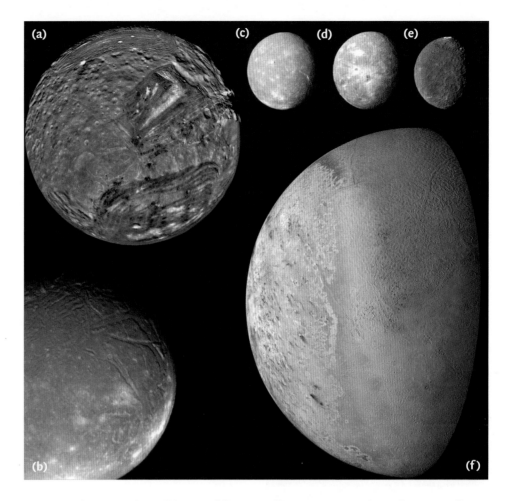

some kind are clearly visible. In addition, Callisto appears to have deposits of sulfur dioxide in some form on its leading hemisphere. Presumably these deposits result from collisions with charged particles in the Jovian magnetosphere. Evidence of fresh ice on Callisto's surface indicates that this little moon continues to sweep up micrometeorites that pound its surface into a fine powder.

The Saturn system has at least 30 moons, including eight major satellites. Most of these little worlds appear as pinpoints of light through our telescopes. The largest satellite is Titan and it's one of the most intriguing moons orbiting Saturn. Voyager showed it to be a world shrouded with methane clouds. Photochemical smog floats high in its atmosphere, which consists primarily of nitrogen, with a small amount of methane. Its frigid temperature keeps methane and ethane in gaseous and liquid states, and makes water ice as hard as rock.

The thick clouds and haze prevented Voyager from seeing down to Titan's surface, and scientists had to wait until HST's Wide-Field and Planetary Camera 2 was installed to take a near-infrared look under the clouds. They took advantage of the fact that Titan's smog layer is transparent to near-infrared light and mapped surface features according to how much light they reflected. The image resolution of this technique allowed the researchers to see objects as small as 576 km across. One large bright area appeared to be just over 4000 km wide (about the size of Australia) and represents some sort of solid surface area. Other bright and dark

Figure 3.34. A series of Hubble Space Telescope images, captured from 1996 to 2000, show the changing viewing angle of Saturn's rings from just past edge-on to nearly fully open as it moves from autumn towards winter in its northern hemisphere. Saturn's equator is tilted relative to its orbit by 27 degrees, slightly different from the 23-degree tilt of the Earth. As Saturn moves along its orbit in its 30-year trip around the Sun, first one hemisphere, then the other is tilted towards the Sun. This cyclical change causes seasons on Saturn, just as the orientation of Earth's tilt causes seasons on our planet. The first image in this sequence was taken soon after the autumnal equinox in Saturn's northern hemisphere (which is the same as the spring equinox in its southern hemisphere). By the final image in the sequence the tilt is nearing its extreme, or winter solstice in the northern hemisphere (summer solstice in the southern hemisphere).

regions could be continents, oceans, craters, or different kinds of surface features.

The rest of the Saturnian moons are cratered, cracked, and in some places are covered with a dark substance that one scientist labeled a "complex organic goo." It's probably the result of bombardment of the surface ices by a constant rain of charged particles in the all-encompassing magnetic field of Saturn. One moon – Mimas – was described as the "Death Star" of Saturn's moons because of its resemblance to the space station from the movie *Star Wars*. A huge crater covers nearly one-third of the surface and if the impactor that created it had been any larger, Mimas might well be scattered in bits and pieces throughout the ring system!

The other moons of Saturn appear to be frigid, icy worlds. Voyager 2 images showed their surfaces as cratered, cracked, and somehow smudged up with dark material – intriguing places for longer follow-up studies by the Cassini mission. Unfortunately, no such follow-up missions are scheduled to study the ice moons in orbit around Uranus and Neptune. That's too bad, because these are among some of the more unusual satellites out there. Uranus has a retinue of five icy moons and a series of smaller satellites. Its larger moons are Ariel, Umbriel, Miranda, Titania, and Oberon. Like the other icy bodies out in trans-Jovian space, they are littered with impact craters. Their surfaces are cracked from the stress of gravitational pull on the brittle ice crusts. Ariel is the brightest of the moons, while Umbriel is the darkest. Titania is scarred with huge canyons. Miranda has so many different kinds of surface features that it looks like a world put together by a committee. Oberon lies the farthest from the planet and like Umbriel, it looks like it has been covered in dusty soot.

When it comes to moons, things aren't any less strange out at Neptune. Triton, as we discussed earlier, spouts geysers from its surface into a thin nitrogen atmosphere. The surface appears to be strangely mottled, and it reminded one Voyager researcher of the surface of a cantaloupe.

Along with retinues of moons, the gas giant planets also boast extensive ring systems. Few sights in the solar system compare to the beauty of Saturn's rings. They have been visible to ground-based astronomers since Galileo's time. They so puzzled him at first that he was at a loss for words to describe what looked like "handles" on the planet. Indeed, the true nature of the rings kept astronomers conjecturing for years. Were they solid or made up of particles? Did they rotate? If so, how fast? If they were solid, how did they manage to stay together as they rotated?

Galileo would have been surprised to find out that his discovery of rings at Saturn was a taste of things to come at the other gas giants. Today, much ring research is focused on the dynamic forces that produced them in the first place. Some scientists surmise that Earth itself may well have sported a ring system early in its formation – possibly as part of the sweeping-up process that built the planet, or the result of encroachments by smaller planetesimals that wandered too close and were broken up by the tidal pull of Earth's gravity. There are two theories about the source of Saturn's ring materials. Either a larger satellite was torn apart by gravitational forces when it got too close to Saturn, or the particles are really

Figure 3.35. Hubble Space Telescope snapshots of Saturn with its rings barely visible. Normally, astronomers see Saturn with its rings tilted. Earth was almost in the plane of Saturn's rings, thus the rings appear edge-on. In the top view, taken in August 1995, Saturn's largest moon, Titan, is casting a shadow on Saturn. Titan's atmosphere is a dark brown haze. The other moons appear white because of their bright, icy surfaces. Four moons – from left to right, Mimas, Tethys, Janus, and Enceladus – are clustered around the edge of Saturn's rings on the right. The rings also are casting a shadow on Saturn. The lower image was taken a few months later, in November 1995, and shows Saturn with its rings slightly tilted. The moon Dione lies at the lower right of the planet. The moon on the upper left of Saturn is Tethys.

material that never actually "got it together" to make a moon when the system was first formed. Regardless of how they formed, the system of rings – and the embedded moons that work to sculpt them and keep them in place – are fascinatingly beautiful.

Although Saturn's rings have been photographed for years using ground-based telescopes, the Voyager 2 spacecraft cameras supplied the best images. It was from these dozens of high-resolution photographs that astronomers first learned that the major rings are actually composed of hundreds of thousands of narrow ringlets. Even such empty-looking regions like the Cassini and Encke Divisions contained faint ringlets. The primary pieces of these rings are ice particles ranging in size from a few centimeters up to several meters across. The total mass of material is spread out across hundreds of thousands of kilometers of Saturnian space, and the ring are anywhere from 10 m to possibly as much as 1000 km thick.

The Voyager missions also revealed some interesting structure within the rings themselves. First, the rings aren't exactly circular, and some of the narrower ones appear to be braided around one another. Something had to be causing these kinky, braided, oblong rings – but what? The culprits turned out to be "shepherding satellites," which orbited just outside the various rings, herding the particles into paths like wayward sheep.

The ring system was the focus of attention in 1995 due to a phenomenon that occurs every 15 years called "ring-plane" crossing. As Saturn and Earth orbit the sun, there are periods of time when the rings appear edge-on. In addition to measuring the starlight shining through the rings, planetary scientists wanted to use HST's high-resolution camera to measure the thickness of the rings, study the

Figure 3.36. **(a)** A Voyager 2 image of Saturn's rings, showing their complex structure. To create this false-color image, Voyager scientists used special filters to screen the rings for possible variations in chemical composition from one part of the system to another. **(b)** Voyager's farewell to Saturn. This picture was taken from a distance of 3.4 million km, and shows some of Saturn's night-side hemisphere.

(a)

(b)

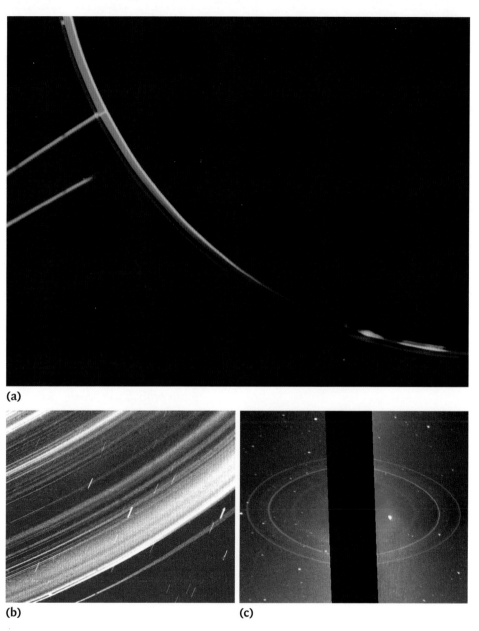

(a)

(b)   **(c)**

Figure 3.37. **(a)** The thin Jovian ring system as imaged by the Voyager 2 spacecraft. Only Jupiter's limb can be seen and the rings are the two shallow arcs stretching out from the planet. **(b and c)** The rings of Uranus and Neptune, also imaged by Voyager 2.

atmosphere, look for the known satellites of Saturn within the rings and find any satellites that Voyagers 1 and 2 may not have seen. Their persistence was rewarded with images taken May 22, 1995 that show four new moons, orbiting near Saturn's eccentric "F" ring. Theoretically all of Saturn's ring particles could be thought of as "moons," although by convention, only the larger ones have been accorded the title.

Not to be outdone by Saturn's glories, Jupiter, Uranus, and Neptune also are encircled by ring systems. Jupiter's is very thin, probably made mostly of dust. It was discovered in an image returned by the Voyager 2 spacecraft, and mapped in great detail by the Galileo mission. The ring has three major components. The largest is the main ring, a dusty band that is about 7000 km wide. It stretches out past the orbits of the Jovian moons Adrastea and Metis and ends abruptly 129 000 km from the center of the planet. At the inner edge of the main ring is the halo, a broad faint torus of material about 20 000 km thick and extending halfway from the main ring to the cloud tops of Jupiter. Just outside the main ring is a pair of faint gossamer-thin rings bounded by the moons Amalthea and Thebe.

The ring systems at Uranus (page 85) and Neptune are comprised of icy boulders mixed with broad dust bands. At Uranus, the larger particles are extremely dark – and researchers think they are coated with some sort of dark carbon-type material.

What about Pluto?

Pluto is the ninth planet, and so far it is the only one that has not been visited by a spacecraft. It is considered a "weird" little place, and – with its companion Charon – is often referred to as a double planet. It has also been in the news in recent years due to several well-publicized attempts to demote it from its exalted status as a planet. It's not larger than some of the moons of the solar system, but it has recently been referred to as the "King of the Kuiper Belt Objects" because it is part of the Kuiper Belt that stretches out from the orbit of Neptune.

Pluto probably deserves its deviant, peculiar reputation. It orbits 5.9 billion km from the Sun in a tilted, highly elliptical orbit. Pluto's "year" is 248 Earth-years long, so only a fraction of one Pluto "year" has passed since Clyde Tombaugh discovered it in 1930. Its companion Charon orbits Pluto every 6.4 days at a distance of 20 000 km. Pluto is twice the diameter of Charon (2300 km vs. 1200 km).

Hubble Space Telescope has studied this distant world, and recent ground-based studies have added a few precious images to the scarce gallery of Pluto pictures. HST imaged most of Pluto's surface in mid 1994. Impact craters or basins could cause variations in bright and dark patterns across Pluto's surface. However, it's quite likely that the planet's thin nitrogen–methane atmosphere is freezing out and depositing complex patterns of icy frosts across the surface of Pluto. As the planet heads into its 60-year-long winter, those patterns could intensify as more of the atmosphere condenses as frost.

Because Pluto is so distant, nobody knew Charon even existed until its discovery in 1979. It was not until HST looked at the pair in 1990 that astronomers were able to clearly separate both components of this "double planet." Now that precise measurements of the orbits can be established with HST and the new generation of ground-based adaptive-optics telescopes, astronomers can refine their estimates of the masses and densities of these two distant worlds. Astronomers think Pluto is about 60 percent rock, whereas Charon comprises mostly ice.

Some astronomers question whether the pair evolved together or may have been thrown together by cosmic circumstance when the solar system was forming. It is possible that they were created, along with other similar "planetary embryos," in the outer fringes of the primordial solar nebula. The other embryos were either used up in the creation of the gas giant planets, or expelled out of the cloud.

Figure 3.38. Pluto and its moon Charon are shown in this sequence of four infrared images obtained on successive nights by the ground-based Gemini telescope using adaptive optics to sharpen the view.

Figure 3.39. Quaoar is the largest object discovered in the outermost solar system since Pluto. This icy world is approximately half the size of Pluto and lies more than 1.6 billion km farther away than Pluto. Both worlds dwell in the Kuiper Belt, an icy belt of comet-like bodies extending billions of kilometers beyond Neptune's orbit. Hubble Space Telescope was able to capture an image of this distant world **(a)**, prompting a space artist to depict it as an icy world with a scarred and darkened surface **(b)**.

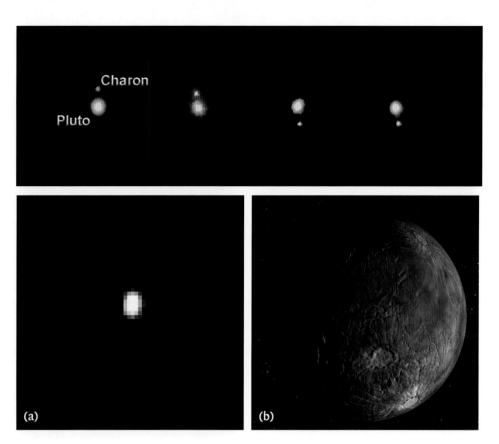

Pluto and Charon may be the only large leftovers. If this is the case, Pluto's "King of the Kuiper Belt Objects" nickname may be as important as its title as planet.

Many Kuiper Belt Objects have been observed beyond the orbit of Pluto. The largest of the finds, dubbed Quaoar (and pronounced *kwa-whar*), lies about 6.4 billion km from Earth. It is almost half the size of Pluto, contains the volume of all the asteroids put together, and probably is made mostly of ice and rock – very similar to a giant comet. Possible future missions to Pluto and the Kuiper Belt may reveal more of these icy objects and their origins. Until then, continuing observations of Pluto with HST and ground-based instruments will supply mission planners with a wealth of information as they decide how best to observe Pluto and Charon up close.

Comets: The windsocks of the solar system

The distant reaches of the solar system are the spawning ground for nearly all of the comets that eventually make their way to the inner solar system. When comets venture close to the Sun they become a perennial favorite of sky-watchers. They're also a nice way to end our discussion of solar system changes. These bits of ice and dirt undergo the most amazing transformations, depending on where they are in relation to the Sun. These icy bodies spend much of their time locked in the deep-freezes of the outer solar system – the Oort Cloud and the Kuiper Belt. Because they formed at the same time as the rest of the solar system did, they are also tracers of our past. They contain chemical mixtures prevalent when the planets were forming, some 4.5 billion years ago. The Oort Cloud of cometary objects is the source of most long-period comets, while the Kuiper Belt appears to

Figure 3.40. A schematic of the Oort Cloud that engulfs the solar system. Embedded within it is the Kuiper Belt – a mass of icy objects that extends from the orbit of Neptune out along the plane of the solar system – and home to the several recently discovered objects like Quaoar, and 1998 WW31 (which appears to have its own moon).

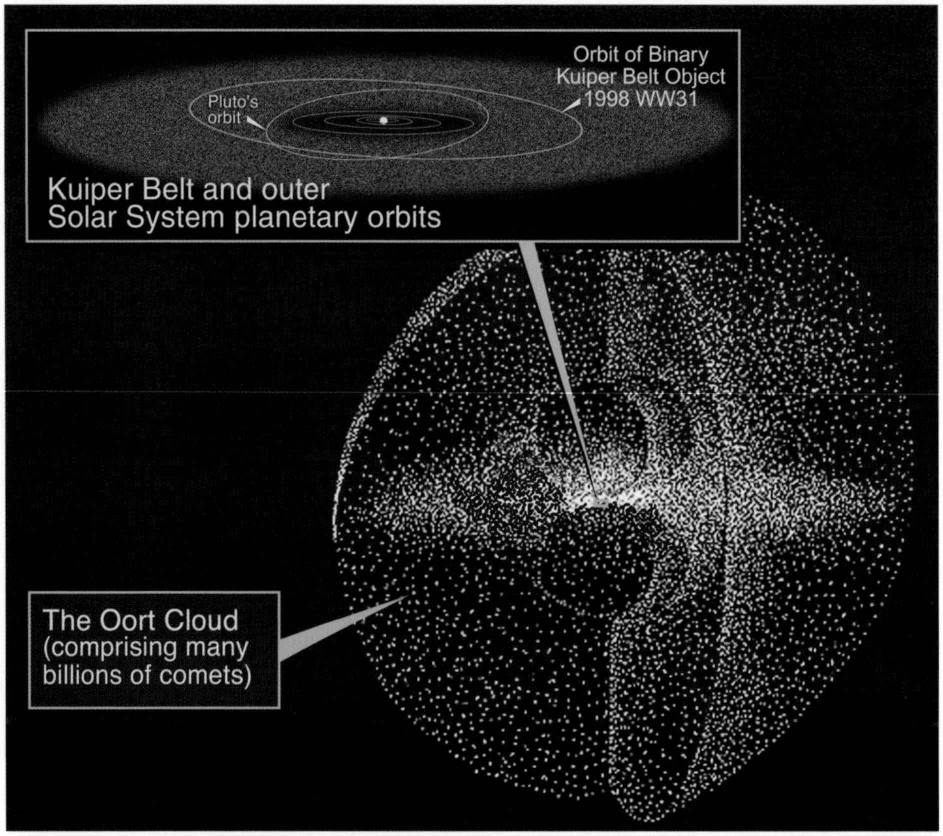

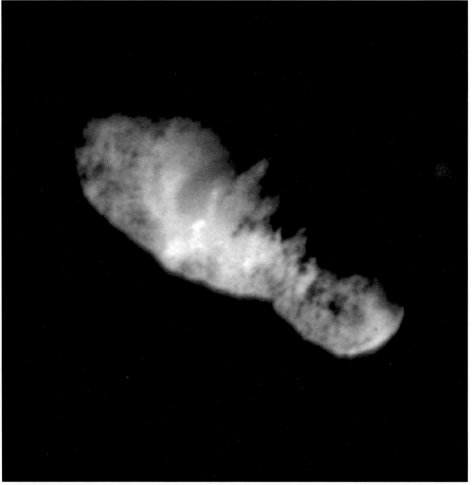

Orbit of Binary
Kuiper Belt Object
1998 WW31

Pluto's
orbit

Kuiper Belt and outer
Solar System planetary orbits

The Oort Cloud
(comprising many
billions of comets)

Figure 3.41. The Deep Space 1 mission passed within 2200 km of the 8-km-long nucleus of Comet Borrelly in September 2001 and found a more rugged terrain than scientists expected. Deep fractures and a dark surface material mar the smooth, rolling plains.

Figure 3.42. On July 14, 2000 the Chandra X-ray Observatory imaged and detected x rays from oxygen and nitrogen ions in the cloud surrounding the comet's nucleus.

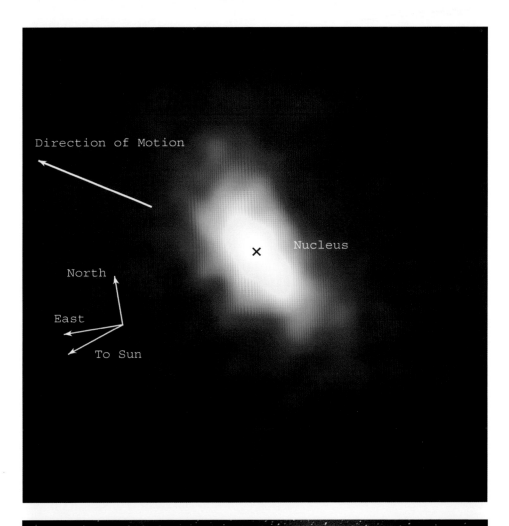

Figure 3.43. Comet C/1995 O1, better known as Comet Hale–Bopp after the men who discovered it – Alan Hale and Tom Bopp – is one of the most spectacular comets of recent history. It came from the far reaches of the solar system and made its last appearance in Earth's skies some 4200 years ago. Its orbit is changing, which means that its next trip past Earth will be in about 2300 years. Hale–Bopp was a bonanza for comet researchers and amateur astronomers, sporting a brilliant dust tail (white and printing straight up in the image) and a strong plasma tail (blue–green, streaming off to the left) during the months of its apparition in 1996.

be the origin of the short-period comets. It is only when comets are tugged from their orbits in the outer solar system into the relatively warmer climes of the inner solar system that they are heated. They sublimate (much as dry ice does) and as they do, they give off the water vapor, carbon dioxide, carbon monoxide, and many other gases that provide clues to their makeup.

The nuclei of comets were a mystery until spacecraft flybys allowed scientists to study their surfaces in detail. The Giotto mission to Comet Halley was the first to get really close to a comet, and it mapped a lumpy, jet-ridden surface. This makes sense, since outgassing is basically what all comets do when they get near the Sun. More recently the Deep Space 1 probe made a flyby of Comet Borrelly and mapped the bottle-shaped nucleus.

The other thing many comets do is grow a pair of tails. One is composed of dust particles and this is the tail we normally see reflecting yellow-white light when we look at a comet. The other tail normally is better seen in blue wavelengths of light. This is the plasma tail, made up of charged particles streaming off the comet along magnetic field lines captured from the solar wind. In fact, plasma tails have long been known as "solar windsocks" for their ability to reflect the properties of the solar wind environment they are experiencing.

Until recently, nobody one expected anything more energetic from comets beyond a scintillating plasma tail and a bright dust tail. That is, no one did until somebody thought to look at a comet in x-ray wavelengths. Normally some very energetic processes in the universe – things like supernova explosions and the annihilation of matter as it enters a black hole give off x rays. But, there they were – x rays being given off by Comet Hyakutake. Astronomers then used the Chandra X-Ray Observatory to look at Comet Linear C/1999S4 and sure enough, it too was giving off x rays. It seems that comets have been giving off x rays all along. As a comet surfs through the solar wind, charged particles (ions) from the Sun blow past. They attract electrons from molecules in the cloud of gases surrounding the nucleus. In the process, the ions and molecules exchange charge and when they do, x rays are produced in a place where scientists least expected them.

As we noted at the beginning of the chapter, astronomy is a humbling pursuit. Occasionally – when we see volcanic flows on Io, or cometary x rays, or evidence of water on dry and dusty Mars – it reaches out and confronts us with a reminder that our place in the universe is far from mundane. Just when we think we know everything, something we thought we understood pretty well reaches out and tweaks us on our complacent noses – telling us that there's more than meets the eye (or detector) in every corner of the cosmos. In the next chapter we'll gaze into space from the comfort of our still exciting and often-mysterious solar system – and see if the stars will surprise us too!

The sky is the ultimate art gallery just above us.
Ralph Waldo Emerson

The oldest picture book in our possession is the midnight sky.

E.W. Maunder

The lives of stars

Gazing into space from the comfort of a temperate planet and exploring the stars and galaxies is an activity that changes a person's life in immeasurable ways. Only the most self-centered among us could look at the sky and not be moved by the beauty that lies out there. The stars have served human needs at one time or another, used for political, religious, and scientific purposes. But only the last really matters if we are to understand the universe in which we live.

The stars live and die on timescales that make the longest human lifespan seem like only a fleeting moment in time. Each one is a unique variation on a basic recipe – a self-luminous ball of gases – that we find cooked up everywhere in the universe. There are hot supergiants, red giants, and Sun-like stars, and they all shine by consuming fuel in gigantic nuclear furnaces at their cores. Some stars travel the galaxy alone, like our Sun. Others perform complex orbital dances around each other as binary pairs, or triplets. Often we find stars in clusters, the largest of which are the so-called "globular clusters," collections of hundreds of thousands of stars packed tightly together in a space less than a few hundred light-years across. No matter what their neighborhoods or how long their lives, however, all stars go through the processes of birth and death.

So, if we look hard enough, we can find the newborns interspersed between the known stars, still nestled in the clouds that gave them birth. We also find stellar geriatrics, some wrapped in death shrouds, or flinging their atmospheres across space in their death agonies, and still others fading into obscurity as low-mass, low-temperature red dwarfs. This chapter focuses largely on those two endpoints of stellar evolution – star birth and star death – because they offer some supremely beautiful visions of the cosmos.

It may seem surprising, but astronomers have only recently developed the tools to interpret what the stars have to say to us. Looking at a sky full of stars is like wandering through a forest and seeing trees of all different shapes, sizes, ages,

Figure 4.1. Star-birth nurseries are usually found in giant, glowing clouds of hot hydrogen (called HII regions). These are essential pieces of the galactic star-birth puzzle. Here, in an infrared study of the star-forming region SL106 IRS4, radiation from hot young stars at the center of IRS4 lights up the surrounding clouds of hydrogen gas. In response, the nebula gives off a beautiful blue emission. The reddish part of the cloud is a reflection nebula, made as surrounding dust particles directly reflect the light emitted from IRS4. Since this infrared image is extremely sharp, we can see subtle details like ripples, filaments, and streamers in the clouds.

Figure 4.2. RCW 108 is another variation on the theme of star formation. Hidden within the clouds near the darkest part of the nebulosity is a cluster of young stars that are in the process of forming. Intense radiation from hot, massive young stars near the center of the image is eating away at the cloud. The bright, red color of the nebula is light given off by clouds of neutral hydrogen.

and states of health. There are tall trees next to short trees, and a forest visitor might decide that tall trees are older than short trees. That could be an erroneous conclusion since there are trees of different species growing side by side, and some short ones might be older than their taller neighbors. Seedlings sprouting up here and there might be mistaken for weeds instead of baby trees. Dead logs and dried-out branches on the forest floor come from the surrounding trees, but an inexperienced observer might not be able to link specific branches to the surrounding trees, or determine why the trees and branches died. One

Figure 4.3. The Flame Nebula glows as part of the Orion Molecular Cloud complex some 1500 light-years away. In this infrared image a dense stellar cluster can be seen in the dark lane separating the two halves of the flame. This cluster is probably less than 1 million years old. Circumstellar accretion disks of gas and dust surround many of its newborns. There could be planets forming inside these clouds even as the central stars go through their own stormy infancy. Infrared imaging allows us to peer through some of the birth clouds to see the stellar youths inside. Just to its south lies NGC 2023, another molecular cloud being eaten away by the heat of massive nearby stars.

might walk across piles of leaves and figure out that some come from certain trees, but how would you explain pine needles? What do pine cones signify? Or acorns?

So, to understand the forest, we have to learn about trees and their life cycles. It is the same way with stars, but unfortunately, humans can't watch the cycle of star life in a linear fashion from birth to death. Our lives just aren't long enough. So, we do the next best thing – we make repeated observations of different kinds of stars at different times in their lives, and come up with theories to describe what we see.

Figure 4.4. The Sagittarius star cloud at the heart of our Milky Way as seen by the Hubble Space Telescope. The view of our galaxy toward the center is obscured by dust, but there are a few open windows through which telescopes can peer. Some of these stellar gems are among the oldest inhabitants of our galaxy. Their colors give clues about their temperature, which allows scientists to make conclusions about their ages and masses. Almost all blue stars are young and up to 10 times hotter than our Sun. They consume their fuel much faster and live shorter lives than our Sun. Red stars can be divided into two types – small stars and red giants. Smaller red stars are usually about half as hot as the Sun and consume their fuel slowly. They have the longest stellar lifetimes. Red giants are at the end of their lives because they have used up most of their nuclear fuel. Many red giants were once ordinary stars like our Sun that swelled up in size as they began their slow spiral to old age and eventual death.

Figure 4.5. Exposures from the Hubble Space Telescope were used to make this composite image of the Large Magellanic Cloud starfield in both visible and ultraviolet wavelengths in a stellar census that revealed more than 10 000 stars covering an area about 130 light-years across. Silhouetted against stars and sheets of glowing gas are dark patches of interstellar gas and dust. The faintest stars in this picture are 100 million times dimmer than the human eye can see. The chevron shape of the image is due to the detector configuration of the Wide-Field and Planetary Camera 2.

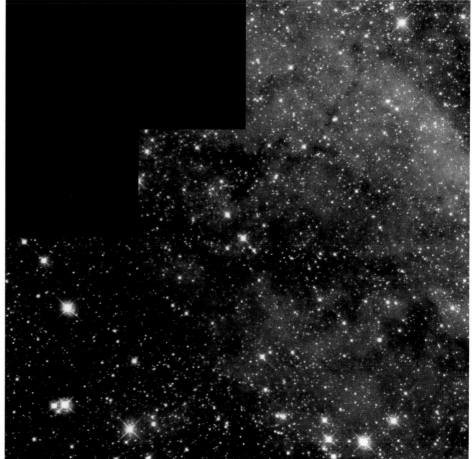

Figure 4.6. The bright cluster Hodge 301, seen in the lower right hand corner of this image, lives inside the Tarantula Nebula, which itself is a knot of stars and gas in our galactic neighbor, the Large Magellanic Cloud. This cluster is not the brightest, or youngest, or most populous one in the nebula, but it is almost 10 times older than other clusters that lie nearby. Many of its stars are so old that they have exploded as supernovae, and the resulting cataclysms have blasted material out into surrounding space at speeds of almost 325 km per second. These high-speed plumes are plowing into the surrounding Tarantula Nebula (toward the upper left), compressing the gas into glowing sheets and filaments.

Figure 4.7. This stellar swarm is Messier M80, one of the densest globular star clusters in the Milky Way galaxy. Located about 28 000 light-years from Earth, M80 contains hundreds of thousands of stars, bound together by their mutual gravitational attraction. Every star visible in this image is either more highly evolved than, or in a few rare cases more massive than, our own Sun. Especially obvious are the bright red giants, which have a mass about that of the Sun's, and are nearing the ends of their lives.

Figure 4.8. Study a globular cluster in the infrared and the stars all appear about the same. In the case of this one – 47 Tucanae in the southern hemisphere – the infrared view tells astronomers that these stars are mostly old red giants and a group of giants called asymptotic branch giant stars. A million stars are packed into the confines of a spherical region of space less than 10 light-years across. By contrast, Proxima Centauri – the nearest star to the Sun – lies about 4.1 light-years away.

The Sun

The Sun is the largest, most massive part of our solar system. If you could pile together all of the planets, moons, rings, comets, asteroids, and space dust that comprises our planetary neighborhood, it would only total about 2 percent of the mass of the system. This monstrous ball of gas is the primary source of energy to power the dynamics of planetary heating. The solar wind – which carries a constant stream of charged particles away from the Sun – blows out a sort of protective bubble called the heliosphere and neatly defines the boundaries of our system.

The energy for all this activity comes from a central furnace deep within the Sun. There the tremendous temperatures (around 15 million degrees) and pressures (more than 300 billion times the air pressure on Earth) create the perfect environment for nuclear reactions to occur. Every second the Sun's furnace converts massive amounts of hydrogen to helium. The resulting energy from this process of nuclear fusion radiates out through several layers to the surface, where it is released as heat and light. The entire process takes about a million years.

The Sun's surface temperature, while not as high as at the core, is a toasty 6000 degrees. This hot surface – called the photosphere – is what we see when we look at the Sun. It often sports dark markings called sunspots. The next layer up from the photosphere is a gaseous zone known as the chromosphere. This is a region

Figure 4.9. The best-studied example of a star is our Sun, which makes it a good place to start when studying the general properties of stars. It stands as our all-purpose reference star from which we can measure and rate all the others.

Figure 4.10. Three images of the Sun taken by the Solar and Heliospheric Observatory (SOHO) were combined to show features that emit at different temperatures. The reddish colored regions show gas at 2 million degrees, the green areas are radiating at 1.5 million, and the blue at 1 million. The fuzzy streamers are plasma flowing out along magnetic field lines from coronal holes and structures above sunspot regions.

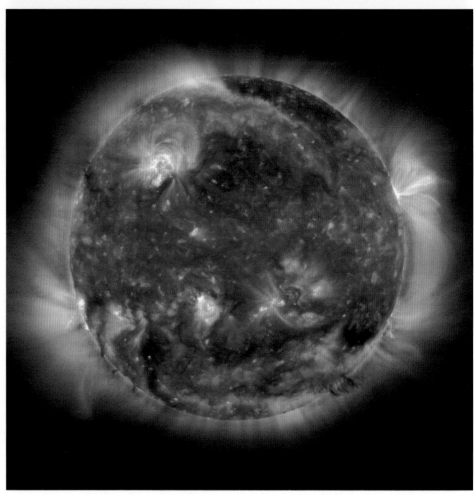

where glowing clouds float above sunspot regions, and tongues of hot gases rise from the surface in the form of flares. The outermost part of the solar atmosphere is the brilliant and exceedingly hot corona. From time to time scientists monitor outbursts from the chromosphere that travel through the corona and away from the Sun as part of the solar wind.

Generally speaking the Sun has been shining for about 4.5 billion years and it probably began to form about 0.5 billion to 1 billion years before its nuclear furnace clicked on. In all that time, the Sun has burned about half its available fuel, and will continue converting hydrogen to helium for at least another 5 billion years. When it exhausts the hydrogen in the central region, the Sun will then start to convert its helium and other elements into usable fuel. It will also start to grow, swelling up large enough to consume the inner planets. It will become a red giant star, and remain that way for a billion years or so before collapsing into a slowly cooling white dwarf star.

Classifying stars

In the grand hierarchy of stars, the Sun comes in comfortably in the middle of the scheme of things. Fortunately for us, our star is not flashy and massive. It's right in the middle of possible stellar characteristics and life cycles. All stars (including the Sun) consume (or have consumed) hydrogen at their cores and radiate the resulting energy. We can take the temperature of a star and measure its brightness (luminosity) – and thus classify it by temperature and luminosity. Temperature usually refers to the temperature at the visible surface of the star. Luminosity is simply a measure of the amount of energy emitted by the star each second. Solar luminosity, for example, is 4×10^{26} watts. Most stars have temperatures ranging from 1000 degrees for the coolest sub-stellar dwarfs to super-hot stars measured at over 200 000 degrees.

Absolute magnitude is a term for the luminosity of a star if it was sitting out at a distance of 10 parsecs away from us. Absolute magnitude puts all stars on an equal footing, brightness-wise, and allows astronomers to focus on the factors like mass, temperature, and pressure that contribute to its intrinsic luminosity. This way stars can be accurately classified by their temperatures and brightnesses.

Astronomers usually refer to other stars in terms of solar luminosity, mass and radius. This gives a nice shorthand way to refer to any stellar characteristics: L_s for luminosity; M_s for mass, and R_s to describe radius. So, for example, if a star's mass is given as 10 M_s, that means it has 10 times the mass of the Sun. The radii of most stars range from about one-tenth that of the Sun (0.1 R_s) to more than 10 solar radii (10 R_s). At the lower end, we are really exploring the cellar in terms of what can be termed a star. To be truly termed a star, an object needs enough mass to produce temperatures that will light up nuclear reactions at its center. Some of the smallest, dimmest things we've observed – largely through infrared studies of the sky – are really "dwarf objects" because they have yet to (and may never) flip the switch on their nuclear furnaces. Amazingly, exotic objects such as neutron stars and black holes also test the definition of the term star, since they are so radically different from their stellar brethren.

When we move up into the supergiant range, stars such as Alpha Orionis (Betelgeuse) are hundreds of solar radii across. At the upper end of the size range, where we see these stars that are 100 R_s, their mass is harder to estimate. At the other end of the spectrum we find white dwarfs packing one solar mass into a volume about the size of the Earth. Neutron stars – those leftovers of supernova

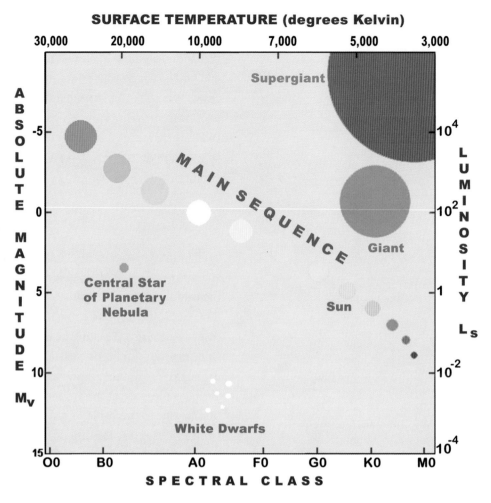

Figure 4.11. Astronomers find it useful to chart stellar characteristics in a well-known chart called the Hertzsprung–Russell diagram (or H–R diagram for short). It works by using stars as points on a plot, with the horizontal axis representing the effective temperature of a star or its spectral class, and the vertical axis representing the star's luminosity or its absolute magnitude. Temperatures can be estimated from the color of the star or by its spectral type. This chart does not necessarily indicate the age of a star, or its distance from us. The colored disks (which are not to scale) give a good feel for the relative sizes of stars – their stellar radii – in the various type categories and at different times in their evolutionary history.

explosions – are thought to be only about 30 km across, but containing several times the mass of the Sun! At the extreme end of tiny we have stellar black holes – tremendous amounts of matter locked forever in a gravitational vault. A black hole of 5 M_s, for example, is a singularity with no radius, but would be surrounded by matter swirling around an event horizon about 30 km across.

Exploring stellar nurseries

When it comes to answering the questions of star birth and star death we find stellar nurseries everywhere we look. Watching the events of a star's birth is a different matter because the process of star formation takes an incredibly long time. The best we can do is to look for as many examples of stars at different points in their formation, and study them in as many wavelengths of light as we can. That process helps us piece together a coherent understanding of the stately progression of star birth – from cold molecular cloud to planetary disks to complex plasma jets and finally to the fiery brilliance of a newly formed star.

The gas and dust clouds in the Orion Nebula are the nearest stellar nursery to Earth. The nebula lies below the three distinctive stars of Orion's Belt and looks, to the naked eye, like a faint greenish haze. Larger telescopes reveal an irregularly shaped cloud scattered with glittering stars. This area of the sky rewards researchers who regularly study it in infrared, visible, and ultraviolet wavelengths.

Figure 4.12. A plot for low-mass stellar objects and substellar dwarfs follows a similar curve to the H–R diagram. Objects classified in the L and T categories are barely visible to the naked eye but are quite evident in infrared wavelengths. Their temperatures are far below even that of a cool M-type star.

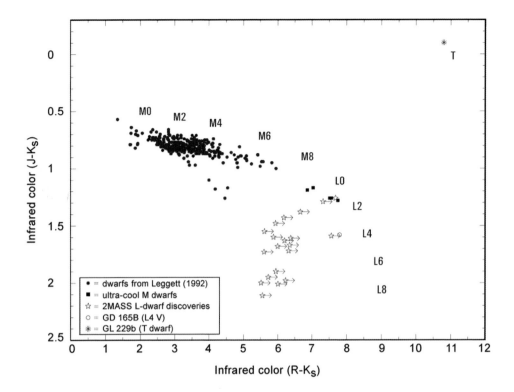

To understand what's happening in the region, it helps to know a little about the interactions of gas clouds that give rise to new stars.

The Orion Nebula is a good example of what astronomers call a blister nebula. It's an illuminated region on the edge of a dark cloud of material called the Orion Molecular Cloud. The blister is the area where the ultraviolet light from young stars is lighting up the surface of the cloud. When we look at this blister, we are seeing a diffuse nebula glowing in the light of hydrogen gas that is heated and ionized by the radiation from bright neighboring stars. Tens of thousands of new stars have formed in this region over the past 10 million years. Some of the newest of the newborns are only a few hundred thousand years old.

The heart of the nebula also conceals a thousand stars that formed about 1 million years ago. These are part of the Trapezium cluster. In visible light, only the brightest of the Trapezium stars can be seen, but in infrared, some incredible details leap out. The search for hot, young stars in the Orion Nebula has led astronomers to study the region in great detail using the infrared detectors on the Hubble Space Telescope and the Very Large Telescope in Chile. What they find helps to answer the question of how something as diffuse and tenuous as an interstellar cloud of gas and dust give can ultimately can give rise to hot young stars.

Star birth is inevitable in interstellar hydrogen clouds. Molecules of gas and dust grains move around and bump into each other in a process called "mixing." Give these cold molecular clouds enough time, and maybe a little gravitational push from an outside influence like a nearby supernova or a passing star, and they will start the process of star formation. A region of higher density starts to form a disk-like shape and material falls toward the center. The cloud spins faster, just as when an ice skater turns faster and faster during a sit spin. At some point,

Figure 4.13. Want to know what our solar system looked like more than 5 billion years ago? Astronomers peered at the dark molecular cloud Barnard 68, using the Very Large Telescope. The image is a color composite of visible and near-infrared images. At these wavelengths, the small cloud is completely opaque because of the obscuring effect of dust particles in its interior.

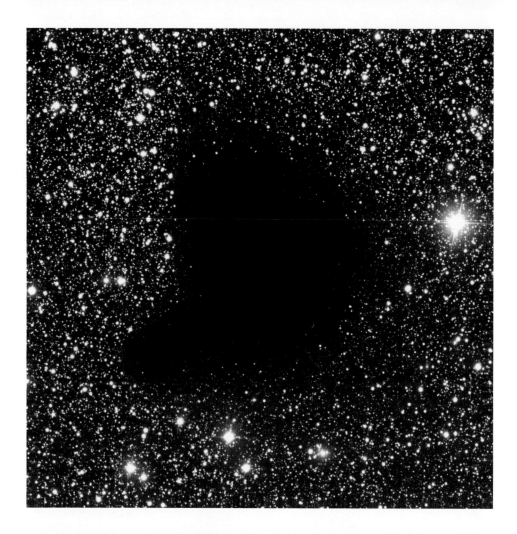

Figure 4.14. The Orion Nebula (M42) stellar nursery is about 1500 light-years away. It is probably the most photographed deep-sky object. The cluster at the heart of the nebula contains several hundred newborn stars, proto-stellar objects, and low-mass dwarf objects. The inner part of the nebula is lit up by radiation from the stars in the Trapezium cluster. The nebula itself is a gaseous bubble formed on the front side of the Orion Molecular Cloud, which also lies about 1500 light-years away. This cloud is the closest site of high-mass star formation to the Sun, and is studied in many wavelengths.

Figure 4.15. If astronomers study the Orion Nebula in infrared wavelengths, more details of the bright, extended nebulosity (due to scattered starlight from dust in the nebula and the light emitted by heated hydrogen gas) are seen. To the north of the Orion Nebula is another part of the Orion Molecular Cloud and a series of red star-like points signifying young, embedded proto-stellar objects.

Figure 4.16. This color composite mosaic image of the central part of the Orion Nebula is made from 81 images obtained with the infrared multi-mode ISAAC instrument on the European Southern Observatory's Very Large Telescope. The famous Trapezium stars dominate the center of the view, along with an associated cluster of nearly a thousand stars of about 1 million years age each.

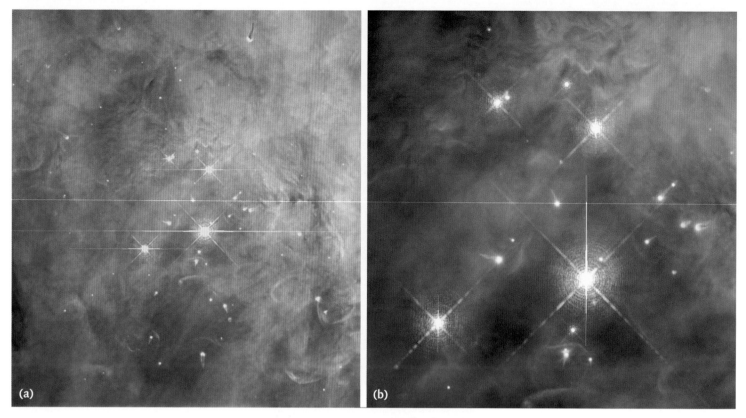

(a)

(b)

Figure 4.17. Two false-color HST images combining observations at several different wavelengths of the center of the Trapezium cluster in M42. The different colors identify the different gases glowing in the cloud. **(a)** In this large-scale view the four massive energetic stars can be seen amid a cluster of evaporating proto-planetary disks. **(b)** A close-up view of Trapezium and its retinue of young stellar birthplaces.

Figure 4.18. (a) This circumstellar disk in the Orion Nebula could easily swallow up our entire solar system. It is a cocoon of gas being eaten away by radiation from a nearby bright star and could be sheltering a newborn star and its planetary system inside. **(b)** The combined winds of star formation in the Orion Nebula sculpt the surrounding clouds of gas and dust into intricate shapes. Here, the Hubble Space Telescope imaged a bow shock around the very young star, LL Orionis (also in the Orion Nebula).

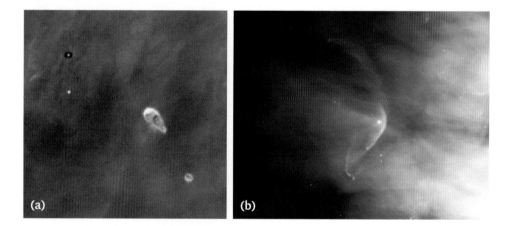

(a)

(b)

the temperature in the center becomes sufficiently high enough for a star to "turn on."

Once the star is born, the leftover material in the disk continues spinning around the star, and occasionally gas from the innermost region around the star is ejected out in a high-speed jet. The protoplanetary disk (also known as a "proplyd") becomes the seedbed for protoplanet formation. The protoplanets accrete material, and over time, create something approximating our solar system. Anything remaining from that process inhabits the farthest reaches of the new

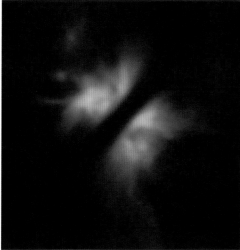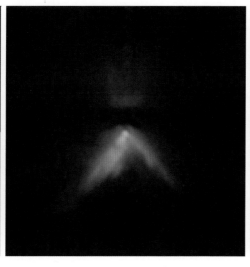

Figure 4.19. The dusty disks surrounding newborn stars (circumstellar disks) are difficult to see in visible light images. Hubble Space Telescope's Near-Infrared Camera and Multi-Object Spectrograph imaged these young stars, immersed in dully glowing clouds that lie about 450 light-years away in the constellation Taurus. Most of the nebulae are made up of small dust particles that reflect light from the stars cocooned within them.

system, as the Kuiper Belt and Oort Cloud of icy objects do for our own solar system.

Now, with all this star birth going on in the universe, one might think that astronomers would see disks around *every* newborn star. They do not, but this does not mean the disks aren't there. These accretion disks are very hard to see because at some stages they are composed mainly of cold gas and dust particles, which are easily outshone by the nearby star. They're usually more detectable in the infrared than in visual or ultraviolet wavelengths.

Herbig–Haro objects

Astronomers have long noticed strange knots of nebulous material that seem to be associated with newly forming stars. George Herbig of the Lick Observatory in California along with Mexican astronomer Guillermo Haro cataloged these objects independently in the early 1950s. At first regarded as curiosities, it turns out that these so-called Herbig–Haro (HH) objects play an important role in star formation. Insight into their contribution to the very early stages of star and planetary system formation has been provided by a powerful combination of detailed, incisive views from HST and wide-field images obtained with sensitive ground-based CCDs. Herbig–Haro objects are produced when bipolar jets of heated plasma stream away from the star and ram into surrounding clouds of gas and dust. As the jet blasts out into the surrounding star-birth nebula, it compresses and heats the clouds, sculpting them into scalloped caverns.

The existence and extent of jets emanating from newly forming stars is a development that astronomers should have expected, but didn't completely figure out until the 1990s. It now makes perfect sense to find them in dusty star nurseries. Here's why: in the accreting nebula, material from the disk falls onto the star and creates hot plasma that blasts back into the interstellar medium via these jets. They take on a narrow, collimated structure as they shoot many billions of kilometers across space from the star. Clearly, some mechanism prevents them from expanding or opening up. Astronomers suspect the interplay between stellar

Figure 4.20. The schematic of star birth simulates a core region about 50 astronomical units wide, illuminated by a protostar (represented by a black dot in the center). Surrounding it is an accretion disk of nebular material, and stretching away from the star are two jets. Given enough time and material, planets would form in the accretion disk.

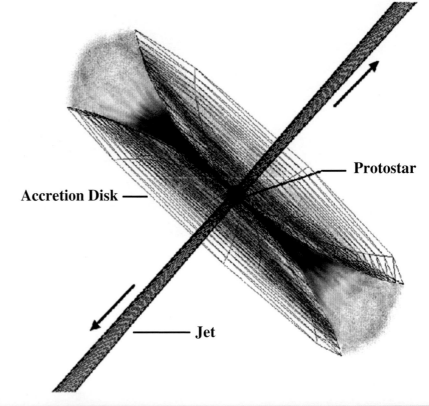

Accretion Disk ——

—— Protostar

—— Jet

Figure 4.21. Herbig–Haro 30 is a real-life example of the schematic in Figure 4.20 It shows a dusty disk edge-on, obscuring the newly forming star (the dust lane between the two bright parts of the disk). The width of the disk is about 12 times the distance from the Sun to Pluto. The jets stretch out perpendicular to the plane of the disk.

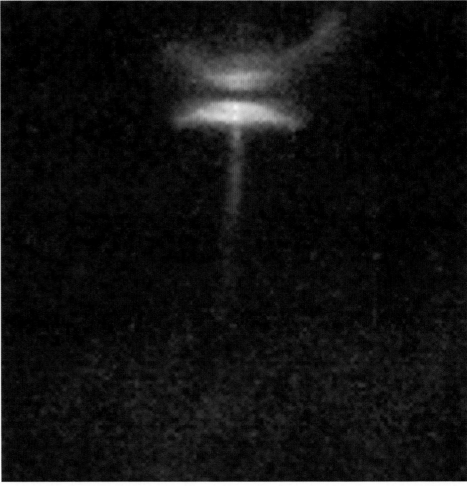

Figure 4.22. **Figure 4.22.** The jet in Herbig–Haro 47 reveals a complex pattern indicating the star (hidden inside the dust cloud near the left edge of the image) could be wobbling, possibly from the pull of an unseen companion star.

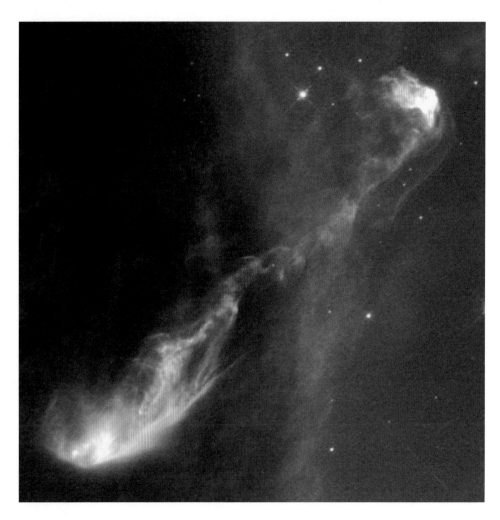

magnetic fields and the inherent fields in the nebula may be corralling the jets into the narrow fields we see.

These recent studies of Herbig–Haro objects provide exceptional insights into the star formation process. Certainly, the outflow of material in jets is an integral part of the star formation process, but a major surprise is the distance to which these jets extend, sometimes across many light-years of space. When we recall that the nearest star to the Sun is just over 4 light-years away, the extent that these jets can influence the star-forming regions of our galaxy is impressive indeed.

Pillars of creation

The Orion Nebula is but the brightest and nearest example of a star-birth region. However, intricately sculpted regions of star formation exist all over the sky. One of the most famous examples is the Eagle Nebula (M16), about 7000 light-years away in the constellation Serpens. In recent years the Eagle Nebula has been observed in infrared detail by the Very Large Telescope array in Chile, and in visible light with HST. Hubble's images show large columns of cold, dense interstellar gas and dust stretching across the sky. The light from nearby young stars illuminates the clouds and the ultraviolet wavelengths are eroding them away. Any stars forming here are still well hidden inside these towering structures.

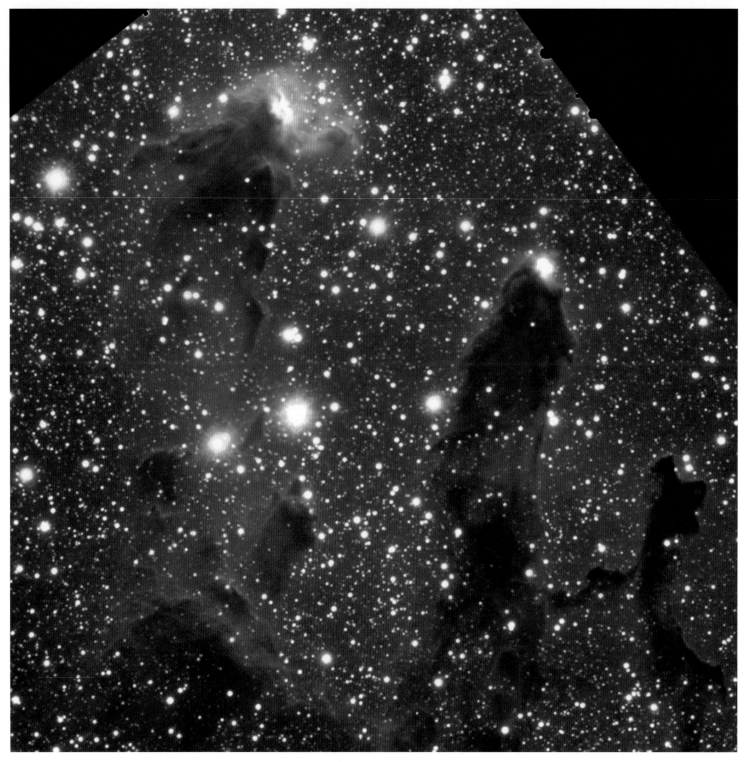

Figure 4.23. Star-birth regions are complex areas where already-formed stars lie side by side with younger siblings that have yet to leave the nest. Astronomers used the Very Large Telescope and a special infrared sensor to zoom into the pillars of the Eagle Nebula. The dusty columns are less prominent than on the Hubble Space Telescope's visible-light image of this same region. This is because near-infrared light penetrates the thinner parts of the gas and dust clouds and only the heads remain opaque. A number of red objects are associated with the pillars: some of these are just background sources seen through the dust, but some are probably young stars embedded in the pillars. The purple arc near the bottom of the picture is Herbig–Haro 216, a fast-moving clump of heated gas emanating from a young star.

Figure 4.24. In the Hubble Space Telescope's visible-light image of the Pillars of Creation in M16, intense light and radiation are eating away at the gas surrounding a number of still-forming star systems embedded in evaporating gas globules. Radiation from a nearby star will eventually eat away the rest of the gas and dust columns, cutting off the fuel to these proto-stellar systems and probably choking their growth completely.

Figure 4.25. This monstrous-looking object is a pillar of gas and dust called the Cone Nebula. It lies in a turbulent star-forming region about 2500 light-years away from Earth. Ultraviolet radiation from nearby stars is eating away at the star-birth cocoon, and causing hydrogen gas to glow a brilliant red. Background stars are peeking through the tendrils of gas. Over time, only the densest regions of the Cone will be left. Inside these regions, stars and planets may form.

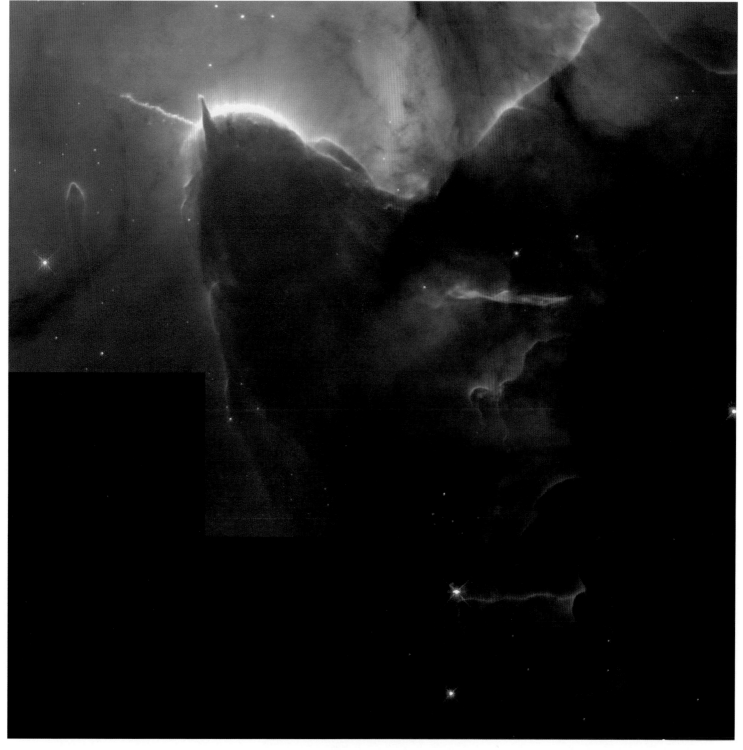

Figure 4.26. What might have happened had the Sun formed in a region crowded with older stellar siblings? The Hubble Space Telescope studied another stellar nursery in the Trifid Nebula where radiation interrupted the star-forming process. Had a more powerful sibling star formed near the Sun, it is very likely that our solar system would not have evolved the way it has. A stellar jet (the thin, wispy object pointing to the upper left out of one of the pillars) is evidence of a newly forming star within the cloud. While this picture is not true color, it is suggestive of what a human eye might see.

The process of evaporation eventually uncovers the young stars and can limit their growth. A young star surrounded by gas and dust and isolated in its birthplace from other stars grows until its mass triggers nuclear reactions in its interior. Then, the star begins to radiate light, and material starts to flow away, first in the form of a narrow jet and eventually a stellar wind. The "mass loss" from a newborn star cleans out the local environment, which effectively stops the star from getting any bigger. If a single star is growing in the proto-stellar cradle, it's far more likely there will be enough material left over to form planets. If enough stars are all forming in the same small region of space, they may choke each other's growth off, and most certainly will use up all the available stellar material in their own formation. In these cases, it is not very likely that planetary systems will form around these stars because there won't be anything left to make even the smallest worlds.

The Eagle Nebula isn't the only place where pillars of gas and dust play host to star formation. The Cone Nebula stretches across 7 light-years of space, and glows in the light of radiation from hot, young stars. Their ultraviolet light also eats away at the cloud, driving gases into the relatively empty region of surrounding space.

Joining the Eagle and the Cone is the Trifid Nebula – home to a stellar nursery that is being torn apart by radiation from a nearby massive star. The star-birth cloud lies about 9000 light-years away in the constellation Sagittarius. At least one newborn star has come to life here, extending a jet out nearly 1 light-year into surrounding space. But, it's a doomed star. Sometime in the next 10 000 years, the radiation from the massive star at the heart of the nebula will finally erode the birth cloud away, choking off the material the star needs to continue growing. Not far from that stellar jet is another stalk pointing out toward the central star from the top of a dense cloud. This "finger" is a good example of an evaporating gas globule. Its tip is lit, possibly from reflected starlight. Radiation from the young star inside may well be trying to escape the confines of its nursery.

Extrasolar planets

Ultimately, the study of star-birth regions brings up the question of planets around other stars. Certainly the dusty disks around newly forming stars are the obvious places to look for planets coalescing in the same way our solar system did 4.5 billion years ago. What people really want to see are already-formed planets orbiting other stars, and international teams are focusing much attention on methods of planetary detection. The Hubble Space Telescope has detected something about 40 to 50 times the size of Jupiter around the nearby star Gliese 229. It's likely this object is a brown dwarf that formed about the same time as its bright companion. More than 100 other worlds have been found orbiting around nearby stars. The numbers are growing as astronomers perfect their methods for searching out dim, distant worlds. The most direct way for finding these extrasolar planets would be to just point a telescope at a likely star system and try to find planets in orbit. The problem with this approach is that planets are

notoriously small and dim – even the Jupiter-sized ones – when compared to stars. They're easily hidden in the glare of their parent stars. So, astronomers rely on more indirect methods that look for the effect that a planet has on a star's light.

One way to do it is by the method of Doppler spectroscopy. When a star has a planet, its orbit is shifted slightly. This causes slight changes in its spectrum. Astronomers call these changes a Doppler shift – a displacement of the spectral lines toward the red or blue end of the spectrum. Planets found using this method are probably Jupiter- or Saturn-size worlds. Earth-size planets would have an effect too subtle to detect.

Another way to find planets is to look at the stars they orbit and see if the stellar brightness changes as the planet passes between us and the star in its orbit. This is known as a transit, and we can observe them in the solar system each time Mercury and Venus pass between us and the Sun. When a planet transits its star, it blocks a very small amount of light, but it's enough for sensitive instruments to pick up. Obviously the bigger the planet, the more light it will block. So, once again, this method works quite well for finding Jupiter-sized worlds, but requires extremely precise measurements for smaller worlds. Another drawback is that the plane of the orbit needs to be in our line of sight for the method to work.

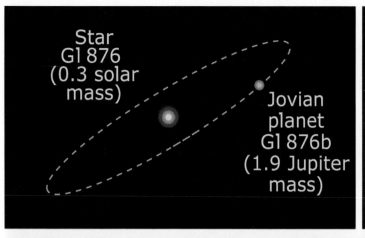

Star
Gl 876
(0.3 solar
mass)

Jovian
planet
Gl 876b
(1.9 Jupiter
mass)

Star Gl 876 without planet: Moves in straight line

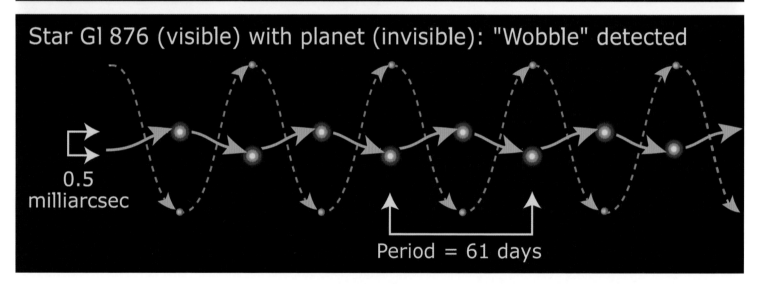

Star Gl 876 (visible) with planet (invisible): "Wobble" detected

0.5
milliarcsec

Period = 61 days

Figure 4.28. The red dwarf star Gliese 876 is known to have something in orbit around it. In 2002, astronomers used the Fine Guidance Sensors on Hubble Space Telescope to measure a "side-to-side" variation in the star's orbit. The wobble in Gliese 876's orbit is due to the tug of the unseen companion object, designated Gliese 876b. The object is approximately two times the mass of Jupiter, and could be a planet. The upper right panel is an artist's conception of the possible planet with several moons, and Gliese 876 in the distance. The Fine Guidance Sensors, sometimes called "star trackers," are able to perform detailed astrometric measurements between objects in the sky. Astrometry is concerned with measuring the positions of celestial objects and their apparent motions.

Finally, astronomers can watch for changes in the star's orbit that might indicate a planet is present. As a planet orbits, it gives a little gravitational tug on its star. This forces the star to wobble a bit in its position. It's almost as if the star is dancing around a point in space – sometimes moving one way and at other times moving the other way. The distance of the star and the length of time it takes for a planet to make an orbit limit this method. Only a few planet detections have been confirmed by this method.

The most exciting thing about looking for these planets is the knowledge that other worlds out there do exist and that we can find them. It's only a matter of time before a researcher somewhere finds a candidate Earth-sized planet around a nearby star – and the astronomical community will rush to study its atmosphere for any hint of water, or possibly even life.

Whistling past the stellar graveyards

Star birth, with its penchant for forming planets, and star death (and its threat to destroy worlds) excite astronomers tremendously. Stellar nurseries and graveyards appear everywhere in the universe. How a star dies is largely dependent on the mass it ends with, rather than its original mass. Stars like the Sun spend most of their lives steadily converting hydrogen to helium. When the process ends, the star goes through the red giant phase. The end of nuclear burning signals the end of the giant phase. Then, the star shrinks, leaving behind a ghostly-looking shell of gas surrounding it, called a planetary nebula. Finally, the star cools and contracts and becomes a white dwarf about the size of the Earth.

Stars slightly larger than the Sun – with final masses between $1.4\ M_s$ and $3\ M_s$ – also spend billions of years on the main sequence, converting their hydrogen to helium. When they contract, they can become "neutron stars," so-called because the intense gravity at the core presses the protons and electrons together to form neutrons. Stars with masses greater than $3\ M_s$ do not end as neutron stars. They explode as supernovae, and the remnant of the dying stars contract further to form stellar black holes. The gravity is so strong that no light can escape the hole, although we do see radiation coming from hot material as it falls into the hole.

The Big Bang and the resulting primordial fireball produced the lightest elements: hydrogen, deuterium, helium, and a little lithium. To get the heaviest elements, you need star death. It's the ultimate recycling process in the universe. Massive stars that explode as supernovae forge elements like carbon, oxygen, calcium, and iron in their nuclear furnaces. When they die, they blow all this material out into the interstellar medium. Supernovae go one step further by creating and ejecting essential elements heavier than iron into the universe. Eventually all this material turns up in other stars, planets, and, ultimately, our bodies.

It's a never-ending cycle of star formation out of interstellar gas and dust, and the subsequent redepositing of stellar materials back into the interstellar medium during the various processes of star death. This also has the result of increasing the amount of heavy elements in the universe over time. The material in the Sun and planets, for example, has been processed through at least one other star. Thus we carry the atoms of long-dead stars in our bodies. Tracing the origins of these chemical elements back through the generations of stars that created them is one way of finding our own place in the universe.

Figure 4.29. A look at the nearby interstellar medium around the solar system provides one way to determine what exists out there and what it's made of. Precise measurements of starlight shining through the local interstellar medium (usually obtained using ultraviolet spectra) give a good idea of the elements present. These measurements, taken over a period of time, can also help determine such things as the speed of a molecular cloud through space. In this computer simulation, the solar system (denoted by the small yellow spot, lower right) appears to lie in a slow-moving cloud of warm hydrogen gas that has been passing through our part of the galactic neighborhood and will leave us behind in less than 3000 years. The cloud of hydrogen gas is being propelled by stellar winds and supernova shock waves from the Scorpius–Centaurus association of hot young stars that lies some 400 light-years away. Our little heliosphere is very close to the "back" end of the Local Interstellar Cloud, protected by the cloud. It will be left behind as the cloud departs. In the coordinate system, the directions are: (GE stands for galactic equator, GC is the galactic center, and NGP is north galactic pole. The 0.0 point is at the three-dimensional center of the Local Interstellar Cloud.)

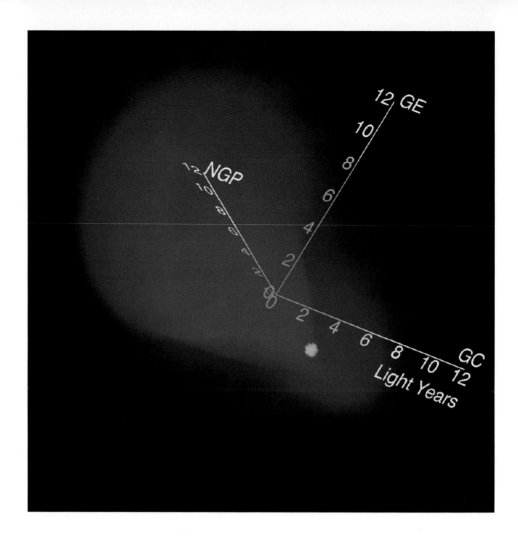

Star death

Most stars spend their lifetimes as so-called main-sequence stars. Once a star gets past the storms of youth, existence on the main sequence is relatively stable, and most of its life is spent happily converting hydrogen into helium. We know that the Sun consumes hydrogen in its core. In the process energy is given off in the form of heat and light. In the early stages of its life, the star is almost all hydrogen. When the hydrogen burns, the "leftovers" or the "ash" is the element helium. Then, as nuclear fusion proceeds, the helium core grows. This stage lasts until about 10 percent of the star's mass has been converted into helium. The Sun, for example, has been on the main sequence for about 4.5 billion years, and should go on for another 5 billion years peacefully converting hydrogen to helium.

Unfortunately, this existence can't last forever. When the helium core becomes so big that the Sun is no longer stable, the fun begins. The core contracts, helium burning commences, and the temperature rises. The outer atmosphere responds by expanding and the Sun becomes a red giant. Eventually it will shrink down to become a long-lived, slowly cooling white dwarf, surrounded for a short time by a cloudy veil of material cast off in earlier stages. A similar fate awaits other stars about the same mass as the Sun, although some white dwarfs with companions can erupt as Type Ia supernovae.

More massive stars than the Sun also spend time on the main sequence. At the end of their hydrogen and helium burning phases, the fusion process continues and creates burning shells of heavier elements – silicon, carbon, and oxygen. In effect, the ashes of one stage are burned as fuel in the next. Finally so much energy is built up in the core that the star swells to become a giant, taking its first step off the main sequence and down the pathway to death.

Eventually a 1 M_S iron core is created. To burn iron would require more energy than would be created. This is when everything comes to a catastrophic standstill. In these very massive stars, the central part collapses, and the outer part of the star blows away in a supernova explosion. What remains will become either a neutron star, or if the progenitor star was supermassive, a black hole.

Planetary nebulae

When stars about the size of the Sun reach their final stage of evolution, they become the central stars of planetary nebulae (so named because through smaller telescopes the nebulae look like disks). Nuclear fusion in the star's core comes to an end. By this time, the core can be a degenerate carbon–oxygen mélange that contains about half the mass of the star. Surrounding the core are hydrogen- and helium-burning shells wrapped in a hydrogen envelope. Stretching away from the star is the huge shell of gas blown off during the previous millennia of mass loss. As the star continues to produce energy, and throw off its outer layers, the hot core is exposed and the surface temperature increases to about 25 000 degrees. This is hot enough to produce a lot of ultraviolet radiation, which ionizes the surrounding clouds of mass-loss driven gas and dust. The shell lights it up, creating a structure

Figure 4.30. The Ring Nebula as imaged **(a)** in visible light by the Hubble Space Telescope, and **(b)** in infrared light by the 2-Micron All-Sky Survey (2MASS). In the 2MASS version, the red color in the outer regions of the nebula arises from infrared emission by atomic and molecular hydrogen. Notice that the central star is very faint in this image, because its high temperature and blue color take it outside the range of the 2MASS instruments.

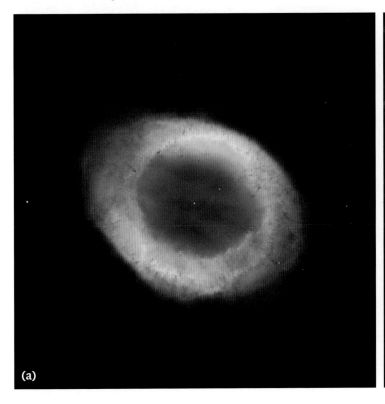

(a)

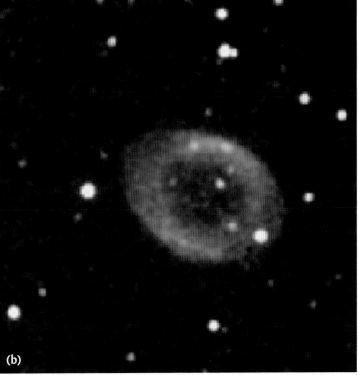

(b)

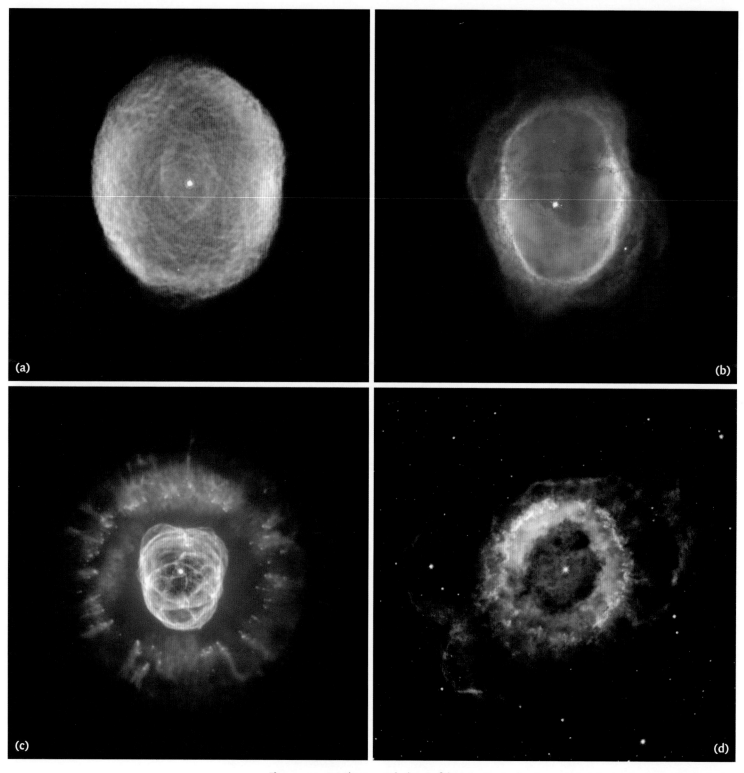

Figure 4.31. **(a)** The star at the heart of the Spirograph planetary nebula (also known as IC 418) was in the red giant phase of its life only a few thousand years ago. It has since ejected its outer layers into space. The stellar remnant we now see is the hot core of the star, which is flooding space with ultraviolet light and causing the cloud to glow. The complex, interwoven texture of the nebula is something of a mystery. Clearly something in the star's environment etched the scalloped lines in the cloud.

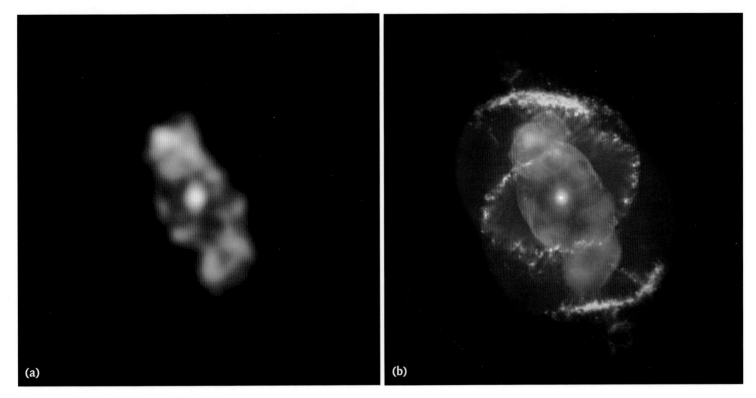

(a)

(b)

Figure 4.32. Peering into the eye of the cat. The Cat's Eye is a planetary nebula that could contain a binary system at its heart. One of the stars is dying and throwing off its atmosphere, and the gravitational effects of the double star's orbital dance create the complex structure. The central star of the Cat's Eye will collapse into a white dwarf in a few million years, but for now its radiation is lighting up the surrounding clouds and causing them to glow and give off x rays. These images **(a)** of Chandra X-Ray Observatons data and **(b)** the famous Hubble Space Telescope image of the Cat's Eye put all the emission areas together in context. The blue areas (shown in orange in the X-ray data) are the hot x-ray-emitting gases, while the red and green structures are much cooler.

of ghostly beauty – the "fossil record" detailing the end stages of the dying star within.

For years, astronomers carried a mental image of the "typical" planetary nebula, based largely on the appearance of the best-known example: the Ring Nebula in Lyra. It fitted the theory of medium-sized star death perfectly, and perhaps satisfied some inner need for symmetry. As astronomers began to find other planetary nebulae, however, the view of the standard planetary nebula began to change. As part of a long-term study of stellar evolution, astronomers have used HST and other high-resolution ground-based telescopes to peer into the hearts of several not-so-symmetrical nebulae, revealing a lovely variety of shapes and asymmetries.

Figure 4.31. **(Opposite) (b)** This expanding cloud of gas, surrounding a dying star, is known to amateur southern-hemisphere amateur astronomers as the "Eight-Burst" or the "Southern Ring" nebula. The expanding gas cloud is nearly half a light-year in diameter, and at its heart are two stars. The bright white one is nearing its own death, but its fainter neighbor is actually the dying star that ejected the nebula we see. **(c)** The planetary nebula nicknamed the "Eskimo" has a double ring structure and glowing wispy threads that look almost like comets with their tails pointing away from the central star. Analysis of the movement of the clouds and gases in the Eskimo suggests that a stellar wind continues to flow from the hot dying star. It began losing mass and forming the outer ring of material about 5000 years ago. The inner ring is only about 1000 years old and is the result of the recent stellar-wind blast. Eventually the nebula will dissipate into space and the central star will settle into a long old age as a white dwarf. **(d)** The Little Ghost Nebula surrounds the remnant stellar core that is sending out a flood of ultraviolet light into the surrounding gas. The prominent blue–green ring, nearly 1 light-year in diameter, outlines a shock front where the ultraviolet light is ionizing the oxygen gas in the clouds. The red light is from nitrogen gas, and blue light is from hydrogen gas. Outside the main body of the nebula, one can see fainter wisps of material that were lost from the star at the beginning of its descent into death.

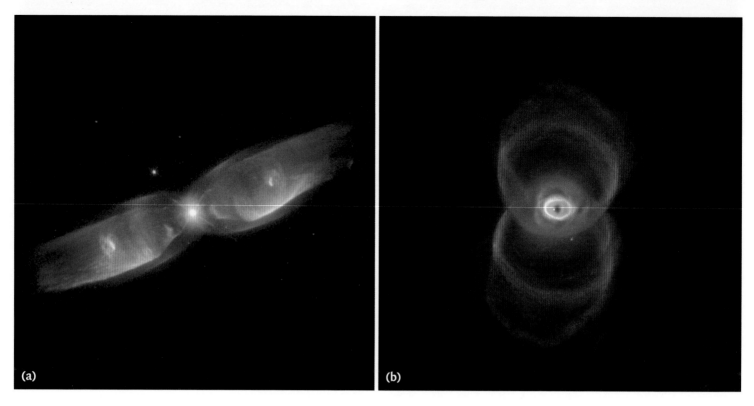

(a)

(b)

Figure 4.33. **(a)** M2-9 is a striking example of a "butterfly" or a bipolar planetary nebula. The central star is one of a very close-orbiting pair of stars. The gravity of one star pulls gas from the other and flings it out into space to create the twin jet "nozzles" we see here. **(b)** MyCn 18, the "Etched Hourglass" nebula, was seen by Hubble Space Telescope in visible light. The complex, symmetrical structure of this nebula may be due to jets of material beaming away from the central star and colliding with gases blown away earlier in the star's evolution.

Many of the planetary nebulae have quite intricate patterns in their shells, leading to a variety of conjectures about gravitational interactions with companion stars to the dying white dwarfs. One of the most complex is the Cat's Eye, in the constellation Draco. It is estimated to be at least 1000 years old. Long observed through ground-based instruments, the nebula was too dim to make out much detail. HST focused on it and found jets of high-speed gas, concentric shells, and shock-induced knots of gas flowing away from the central region at more than 6 million km per hour. At least one theory suggests that the star at the center might be a binary system. A fast stellar wind blowing off the central stars may have created the curious ellipsoidal shape of the inner gas shell. Surrounding it are two larger lobes of gas blown away from the star. Jets of gas that seem to point in different directions may have formed the bright arcs and curly-shaped structures. It's possible that the jets are wobbling as the stars orbit around each other, and that action could be possibly turning the jets on and off like a beacon. While HST was able to see regions of relatively cool gas, the Chandra X-Ray Observatory telescope concentrated its attention on the hot gases in the center of the nebula, recording evidence of collisions between hot and cold gases that produce x rays.

Figure 4.34. The so-called "Ant Nebula" (Menzel 3, or Mz 3) resembles the head and thorax of a garden ant. This dramatic Hubble Space Telescope image reveals the "ant's" body as a pair of fiery lobes protruding from the dying, Sun-like central star. This wispy planetary nebula could be hiding a second, dimmer star orbiting with the bright one that formed the nebula. Interactions could have shaped the flow of material from the dying member of the pair. It's also possible that the dying star's own strong magnetic field is channeling the gas out into space.

Supernovae

Nothing catches the attention of astronomers quite like the final, cataclysmic explosion of a star as a supernova. In their brightness these temporary but spectacular outbursts sometimes rival the luminosity of entire galaxies. Because they play such a pivotal role in stellar evolution, supernovae afford astronomers a chance to study the ongoing enrichment of the interstellar medium in heavy elements, the creation of elements heavier than iron, the formation of neutron stars and black holes, and a probable role in the conception of new stars in nearby HII regions (see Figure 4.1).

The remnants of supernova explosions are scattered across the sky. One of the most extensive and often photographed is the Veil Nebula, also known as the Cygnus Loop. It is an enormous region of diffuse gas emissions. The nebula consists mostly of interstellar matter swept up by the material flung off by the exploding star, and it shines because of excitation due to collisions between this expanding shock wave and the surrounding interstellar medium. The Veil Nebula also emits weak x rays.

The best-known (and earliest observed) example of a supernova remnant is the Crab Nebula in the constellation Taurus. It exploded into view in July 1054 AD and remained visible both day and night for several months before fading into

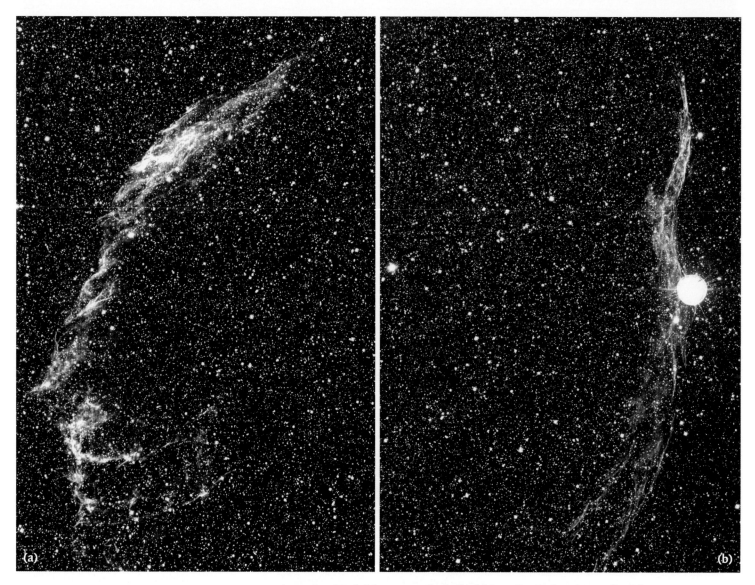

Figure 4.35. The eastern (a) and western (b) halves of the Veil Nebula in the constellation Cygnus. These are supernova remnants from a massive star that exploded more than 20 000 years ago.

obscurity. Had this star exploded within a few light-years of Earth, our system would have experienced increased radiation from the explosion, in addition to energetic particles moving out away from the site at perhaps half the speed of light. Fortunately, the explosion occurred some 6000 light-years away and the solar system probably experienced a brief bombardment of neutrinos rushing away from the site. There are records around the world recording this remarkable stellar event, including notable Chinese and Japanese accounts, where it was noted as a "guest star." There is evidence to suggest that Anasazi observers in what is now the American southwest also noted the star's appearance and painted a record onto a sandstone ledge.

Today the Crab Nebula remains an object of intense interest among astronomers who are studying the leftover remnant of the supermassive star that died in the explosion. The supernova that produced the Crab Nebula formed a pulsar – a rapidly spinning neutron star. Its signal flares out 30 times per second as the star turns and blasts a jet of radiation our direction. Electrons being whirled around in the strong magnetic field of the star probably generate the pulses. As

Figure 4.36. Two visible-light views of the Crab Nebula taken by the Very Large Telescope. **(a)** A wide-field view of the Crab; **(b)** view zeroing in on the pulsar at the nebula's heart. The blue light is emitted by very high-energy electrons that spiral in the large-scale magnetic field. They are accelarated and ejected by the rapidly spinning neutron star (a pulsar) at the center of the nebula. The pulsar is one of two close stars near the geometric center of the nebula.

Figure 4.37. The Crab Nebula in a combined visible and x-ray image from Hubble Space Telescope and the Chandra X-ray Observatory. This image captures a still life of matter propelled to approximately half the speed of light by the Crab pulsar, a rapidly rotating neutron star at the center of the nebula. Bright wisps moving rapidly outward form an expanding ring that is visible in both x-ray and visible images. These wisps appear to originate from a shock wave that shows up as an inner x-ray ring. This ring consists of about two dozen knots that form, brighten and fade, jitter around, and occasionally undergo outbursts that give rise to expanding clouds of particles, but remain in roughly the same location.

these particles are hurled into the surrounding gases by the spinning of the magnetic field, they radiate energy, causing the cloud to glow.

Supernova 1987a

Although supernovae occur in distant galaxies many times a year, it's rare that we see a bright one relatively nearby. Since January 1987, Supernova 1987a in the Large Magellanic Cloud has provided astronomers with a splendid opportunity to observe the death of a star and a spectacular chance to watch the aftermath of a stellar explosion. Since HST first observed it in August 1990, studies have focused on the cloud of debris rushing out from the supernova.

The ring appears too far away from the supernova to consist solely of material that was blown off during the event. The progenitor star most likely experienced extensive, ongoing mass loss in the form of a stellar wind. It was this outgassed cloud that was brightened by the light flash from the supernova some 240 days after the explosion. If this phenomenon follows theory, there should be a pulsar showing up at the heart of supernova 1987a. So far, none has been detected, but it could be just a matter of time before a signal makes its way out from the expanding cloud of debris. It's also possible that the remnant of the star may be too massive (greater than about $3M_S$) to become a neutron star and instead has collapsed to a black hole. A third possibility is that the Earth may simply be out of the line of sight of the pulsar beam and we are just missing seeing it.

Figure 4.38. **(a)** Supernova 1987a (the bright star just below and to right of center) in the Large Magellanic Cloud as seen by the Cerro Tololo Inter-American Observatory in Chile, made three days after the star's outburst.
(b) (Opposite) Several years later, the Hubble Space Telescope imaged the expanding rings of debris being lit up by radiation from the dying star. The hunt is now on for a pulsar at the site of the massive stellar collapse.

(b)

Eta Carinae, the cataclysmic variable star

The Milky Way galaxy has a number of stars called cataclysmic variables, part of a larger class of variable or unstable stars, and we bring our exploration of star death to an end with one that has not quite died yet. This is the luminous blue star Eta Carinae, and one look at it would be enough to convince anyone that this is a star in its death throes. This is a massively unstable and relatively young star that is blasting huge amounts of material into surrounding space. What we see today is a rapidly expanding shell, radiating out in lobe-like structures from the central star. The reddish glow is actually light from fast-moving nitrogen and other gases ejected from the interior of the star at more than 3 million km per hour. The bright white material is very dusty and reflects starlight back to us. Astronomers are still trying to explain the dynamics of Eta Carinae's behavior. Its ultimate fate is to explode as a supernova, but besides its appearance it exhibits few other characteristics that would label it as such. The violence of the outburst of this cataclysmic star leads one to wonder what it will look like when it really does blow up!

Figure 4.39. This visible light image of the Carina Nebula – with Eta Carinae at its heart – is a composite of several exposures and captures the light emitted by ionized (heated) oxygen gas, hydrogen gas, and sulfur. Eta Carinae is the lower left of the two bright stars in the nebulosity near the center.

Figure 4.40. An infrared image of the Eta Carinae nebula. Eta Carinae is the feature at the bottom left of the upper nebulosity. Compare with Figure 4.39.

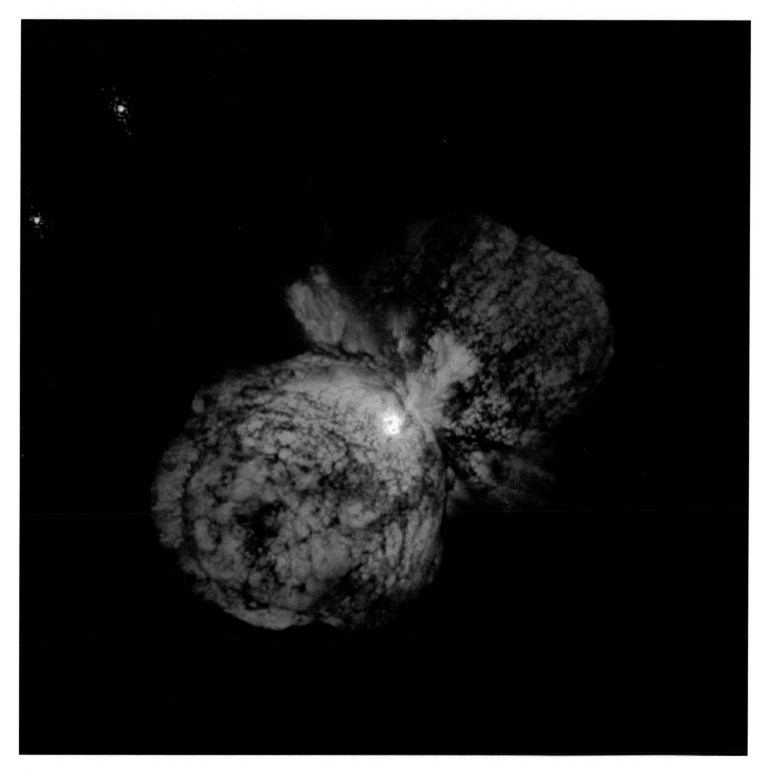

Figure 4.41. Eta Carinae itself as seen by the Hubble Space Telescope. Even though Eta Carinae is more than 8000 light-years away, we can see structures in the expanding cloud that are about the size of our solar system.

The nebula surrounding Eta Carinae contains stellar nurseries that are birthing the next generation of stars. The shock waves generated by this star's passing will almost certainly trigger an even greater rash of stellar births – a wonderful example of the cycles of star death and star birth that reverberate throughout the galaxy.

The mystery of the gamma-ray bursts

Supernovae are not the only bright things popping off out there in the sky. Astronomers have long known about bright gamma-ray bursts and puzzled over what sort of object could radiate that much energy in a short period of time. In chapter 1 we discussed gamma rays as the most energetic radiation known – having at least 100 000 times the energy of visible light. Whatever causes gamma radiation has to be very energetic indeed! Cosmic gamma-ray bursts were first discovered as part of the Vela satellite program to monitor the former Soviet Union for compliance with the nuclear test-ban treaty of the mid 1960s. In recent decades a host of spacecraft missions has swept the sky, looking for the source of these outbursts. These included the Ulysses mission, the Compton Gamma-Ray Observatory, the X-Ray Timing Explorer, and the BeppoSAX satellite.

The data from these missions have allowed astronomers to make a few general conjectures about the source of gamma-ray bursts (or GRBs, as they're known in the community). The big question was whether they were something in our own galaxy – or existed in other galaxies. The accumulated observations amount to be about three bursts per day, coming equally from all directions in the sky. This argues for at least some of the sources to be in other, much more distant galaxies.

Generally the bursts rise to their maximum intensity in about 0.1 second. For something to produce this amount of radiation this fast, it can't be any farther across than the distance light travels in 0.1 second – making the source only a few tens of thousands of kilometers across – something almost unimaginably small, but extraordinarily bright for a very short period of time.

Figure 4.42. Data from the Compton Gamma-Ray Observatory was used to make an all-sky map of bursts. They appear all over the sky, which means they probably emanate from extragalactic sources. The coordinate system refers to galactic longitude and latitude.

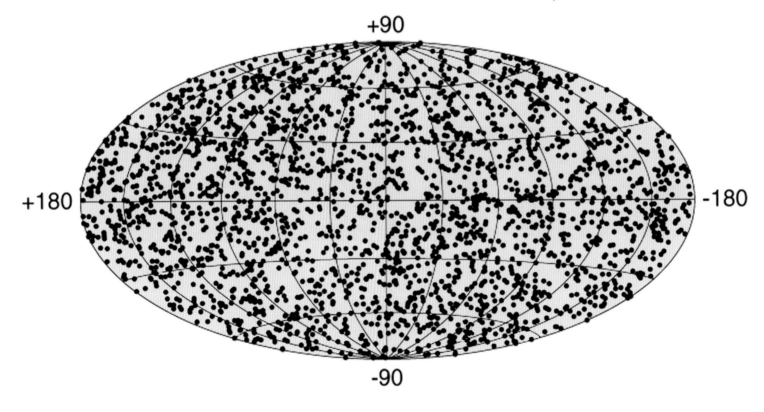

Figure 4.43. The fading afterglow of a gamma-ray burster, as seen by the Hubble Space Telescope in June 1997. Explanations for these distant fireworks explosions range from collisions of neutron stars to some galaxy-related phenomenon.

For a while there was controversy over the location of the bursters, but that too has been resolved with recent data. It turns out they are almost certainly some 3 to 10 billion light-years away. For something that far away, astronomers estimated a gamma-ray burster would have to be an object putting out energy at something like 10^{53} or 10^{54} ergs in all directions (what astronomers call isotropic emission). To give you an idea how much energy this is, consider that the Sun puts out about 4×10^{33} ergs per second. It would have to burn at its present luminosity for just over 7 billion years to equal the energy emitted by an isotropically emitting gamma-ray burster in a fraction of a second! The production of this huge amount of energy means that we are dealing with major fireballs occurring in the universe. There isn't anything that small that could do such a thing. Yet, there they are out there – bursting away at a rate of three a day. Clearly astronomers have to think a bit harder about what these objects could be.

Perhaps the idea of an object that radiates a huge amount of energy out in all directions isn't correct. Maybe a gamma-ray burster is something with a compact central "engine" that pours out a narrow, focused beam at 10^{51} ergs per second. Admittedly that's still a huge amount of energy, but it's more realistic than the numbers we just mentioned, and allows astronomers to consider a couple of types of objects. One idea is that gamma-ray bursters are collisions of neutron stars or black holes. Shoving two such massive objects together could produce prodigious amounts of energy. Another possibility is that all galaxies have some other kind of powerful objects that can briefly radiate huge amounts of energy.

Astronomers are not even close to an explanation, but they continue to monitor the sky for these energetic beasts. Visible wavelength observations give astronomers accurate positions and allow them to follow the evolution of the sources, including the afterglow that fades away after the main event. This type of work is often a collaboration between gamma-ray, x-ray, and optical observations from spacecraft and optical and radio observations from the ground. For example, on February 28, 1997, the Italian–Dutch satellite, BeppoSAX detected a 3-minute-long burst in the direction of the constellation Orion. It was designated as GRB 970228. Naturally, this set off a flurry of activity. HST swung around and obtained images of the fireball in visible wavelengths. A galaxy appeared to be the host for the burster. Subsequent observations tracked the fading glow of the burster.

Another gamma-ray burst was detected by BeppoSAX on May 8, 1997 and imaged by HST's imaging spectrograph on June 2 (Figure 4.43). The fading fireball is seen in the center of the STIS image, but no host galaxy is visible. If there is one, it is much fainter than the Milky Way.

If a gamma-ray burster is a beamed light signal from some sort of cataclysmic event, we may not be seeing all of them because the beams may not always be pointing right at us. But, what we might see would be an orphan afterglow – the fading aftermath of such an event. If this model is correct, a great many bursters may not be aimed directly our way. This means that there could be roughly 500 times more gamma-ray bursters than we actually observe and that orphan afterglows should be more common. So far they're quite rare. The expected rate of gamma-ray bursters is one per galaxy every 100 000 years, so it looks as if the beaming model may need a bit of adjusting to explain these enigmatic outbursts completely.

Gamma-ray bursters and their hidden causes are still very much an astronomical research work in progress. Some researchers now think there may even be two different types of bursters. If they're the result of two merging neutron stars, or something even weirder, things could turn out to be very interesting in our own and other galaxies. As they say on the TV news: "Stay tuned!"

5 Galaxies: Tales of stellar cities

Astronomy is the science of the harmony of infinite expanse.

Lord John Russell

The galaxy hardest for us to see is our own. We are far out from the center . . . we lie in a spiral arm clogged with dust. In other words, we are on a low roof on the outskirts of a city on a foggy day.

Isaac Asimov

Galaxies are stellar cities that wheel through the universe like cosmic snowflakes. No two look exactly alike. A picture of a galaxy is a snapshot that freezes it in a moment of time. And that time is necessarily in the past because the farther we cast our gaze into space, the further back in time we see. The great distance at which galaxies lay works against astronomers as they study the ongoing long-term changes that alter the structures of these stellar collections. With lifetimes that can stretch back to the earliest epochs of the universe, it's also hard to see how galactic structures evolve. So, as with stars, we have to look at many different galaxies at all distances to understand how they form and how they change through time.

The starry swath of the Milky Way gives us our first clues about galaxies – and a unique vantage point. The thought that we are looking edgewise through "our own galaxy" was – and perhaps still is – completely astounding to some. The Milky Way's starry arch across the sky has evoked awe and wonder in the human mind since prehistoric times. Each culture has given it a meaningful name, such as La Via Lactea, the Backbone of Night, the River of Heaven, the Starry Spirit Road, and many others. It was not until Galileo Galilei turned his telescope to the Milky Way that astronomers saw this collection of stars for what it really is. It remained for astronomers in the twentieth century to determine the shape and extent of this river of night.

The words used to describe galaxies through the years illustrate the historical confusion about just what they really were. The term "nebula" used to be applied to any fuzzy looking, cloud-like structure that obviously was not a star. In fact, the eminent eighteenth-century astronomer Charles Messier compiled his famous Messier catalog of "fuzzy-looking things" (which is the origin of the "M" numbers you see applied to some astronomical objects), so that comet-seekers would not be confused by the permanent nebulae in the sky. In the early decades of the twentieth

Figure 5.1. From our viewpoint inside the galaxy, the Milky Way is a swath of stars overpowering the night.

Figure 5.2. A color mosaic of the Milky Way reveals the location of star clusters, gas and dust lanes, and the brightly lit center. The Large and Small Magellanic Clouds (companions to our galaxy) are the two irregular blobs of light to the lower right.

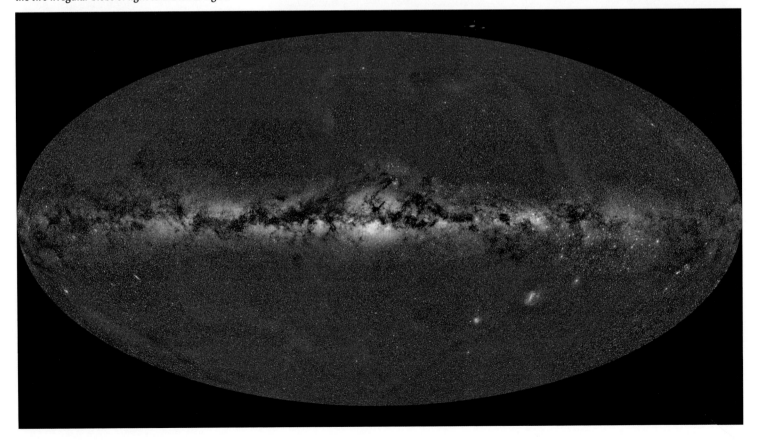

Figure 5.3. A false-color infrared view across 25 000 light-years of space to the center of our Milky Way galaxy uncovers the location of thermally heated clouds of gas and dust around the central core – where a supermassive black hole is probably lurking. If we were to look at this scene in visible light, most of the stars would be hidden behind thick clouds of dust. This obscuring dust becomes increasingly transparent at infrared wavelengths.

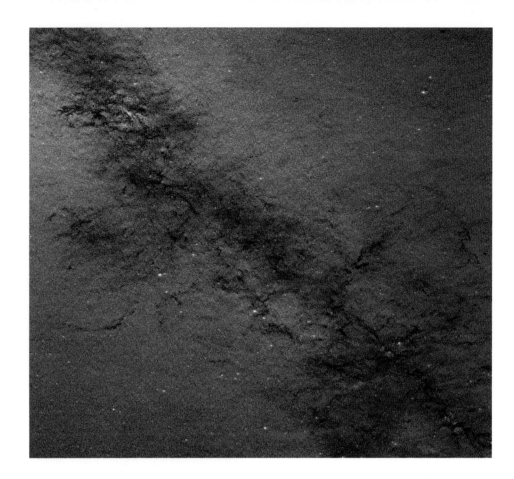

Figure 5.4. The majestic spiral galaxy NGC 4414 lies about 60 million light-years away in the constellation Coma Berenices. The central regions contain primarily older, yellow and red stars. The outer spiral arms are the sites of young, blue stars, the brightest of which can be seen individually. The arms are also very rich in clouds of interstellar dust where future generations of stars will form.

Figure 5.5. The "Sunflower Galaxy" lies 36 million light-years away from Earth in the constellation Canes Venatici (near Ursa Major). Like other spiral galaxies, it is studded with knots of star birth and the gas and dust cradles that harbor stellar nurseries. The image clearly shows the galaxy's tightly wrapped spiral arms and a multitude of small reddish HII regions where hydrogen gas is glowing due to the presence of newly formed hot massive stars.

Figure 5.6. This galaxy in the southern hemisphere constellation Circinus belongs to a class of mostly spiral galaxies that have compact, active centers and are believed to contain massive black holes. The central regions of these galaxies blow superheated gas out into space at phenomenal speeds. Astronomers studying this one are seeing evidence of powerful emissions, which heat up the rest of the galaxy, causing it to glow in visible and infrared wavelengths.

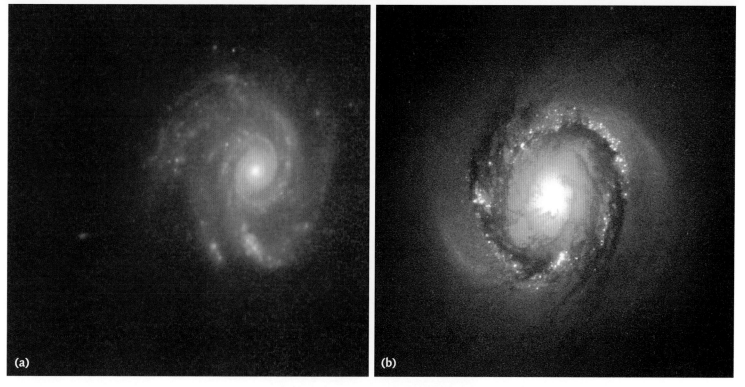

Figure 5.7. **(a)** The dull infrared glow of hydrogen gas clouds in star-birth regions penetrates through the dusty environs around the center of galaxy NGC 5653. The red knots outlining the curving spiral arms in NGC 5653 pinpoint rich star-forming regions where massive newborn stars are pouring out ultraviolet radiation to heat the clouds. Middle-aged stars and stellar clusters show up in white light. The dark material is dust. **(b)** The galaxy NGC 4314 appears in a composite of ultraviolet, blue, visible, infrared, and a specialized hydrogen-alpha filter. The purple color represents hydrogen gas being excited by hot young star clusters. Clusters of infant stars formed in a ring around the core and are the only places where star formation is taking place in this barred spiral galaxy.

Figure 5.8. **(a)** A nearly perfect ring of hot, blue stars pinwheels about the yellow nucleus of an unusual galaxy known as Hoag's Object. The entire galaxy is about 120 000 light-years wide, about the same size as our Milky Way galaxy. The blue ring, which is dominated by clusters of young, massive stars, contrasts sharply with the yellow nucleus of mostly older stars. What appears to be a "gap" separating the two stellar populations probably contains some faint star clusters. **(b)** Could Hoag's Object once have looked like this in the distant past? Two spiral galaxies pass by each other like majestic ships in the night. NGC 2207 (the larger galaxy) is distorting smaller IC 2163 with its strong gravitational pull. Long streamers of stars and gas are stretching out 100 000 light-years from the scene. Astronomers think that this galactic dance will continue far into the future as each galaxy tugs at the other.

Figure 5.9. The galaxy UGC 10214 – nicknamed "The Tadpole" – resides about 420 million light-years away in the constellation Draco. It probably got its distorted shape in a close encounter with the small compact galaxy tangled in the head of the Tadpole. Strong gravitational forces from this galactic tango created a long tail of gas and clusters of newly created stars that stretches out more than 280 000 light-years. These will eventually become old globular clusters similar to those found hovering in the halo of the Milky Way galaxy. The two brightest clumps of stars in the tail could become companion dwarf galaxies. This scene of galactic interaction and its resulting torrent of star birth are playing out against a backdrop of even more distant galaxies.

Figure 5.10. The "Sombrero Galaxy" (so named because it resembles a large, flat-brimmed Mexican hat) lies about 50 million light-years away in the constellation Virgo. Many globular clusters swarm around its well-populated central bulge, and this image reveals its fine spiral structure. We see it nearly edge-on, and even small telescopes can capture its unusual shape.

century, astronomers argued over whether these hazy-looking blobs and so-called "spiral nebulae" were in our galaxy or outside of it. Eventually, with the advent of more powerful telescopes, and spectroscopic techniques to analyze the light coming from these objects, astronomers realized that many dim, fuzzy nebulae were, indeed, outside the Milky Way. The next step was to understand what they were and how they came to be.

Classifying galaxies

Galaxies are everywhere in the universe. They come in a veritable rogue's gallery of shapes, sizes, and masses. Some are quite distant and hard to see without at least a fairly good-sized telescope, but many are accessible to observers with small telescopes, and surprisingly – even the naked eye. The most obvious galaxy is our own – but of course we're seeing it from the inside out. Beyond the Milky Way, there are exactly three galaxies available for naked-eye stargazing. The spiral-shaped Andromeda Galaxy (M31) lies about 2.5 million light-years away (in the constellation Andromeda) and contains several hundred billion stars. It is visible in dark skies well away from light pollution. The other two "naked-eye" galaxies are actually companions to ours – the irregular-shaped Large and Small Magellanic Clouds in the southern hemisphere skies.

Figure 5.11. Stargazers in the northern hemisphere sky have one naked-eye galaxy outside the Milky Way to observe. It's the Andromeda Galaxy, also known as M31, not far from the W-shaped constellation Cassiopeia. It lies about 2.5 million light-years from us and has several hundred billion stars.

Galaxies: Tales of stellar cities

M31 is a typical spiral galaxy. Spirals rotate, taking hundreds of millions of years to make a complete revolution. The spiral arms are of special interest to us for several reasons. First, our solar system bobs around the galaxy about one-third of the way out from the center in the general region of the Milky Way's Orion Arm. Second, spiral arms are prime sites for the birth of stars (the stellar nurseries in the Orion Nebula are a good example). You might say that we have the inside view on life in a spiral arm. Finally, since we know that spirals have many star-formation regions, by searching them out we can use them to look back to a time when clusters of young galaxies (many of them spirals) experienced massive amounts of starburst.

As a general rule, galaxies with more open arms tend to have more gas and dust, and show a greater tendency for brisk episodes of star formation. In the early

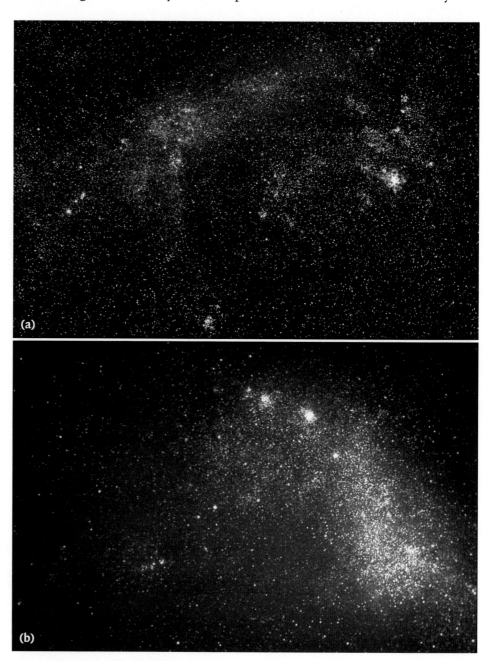

Figure 5.12. The Large (a) and Small (b) Magellanic Clouds in the southern-hemisphere sky. They are a pair of irregular galaxies in the constellations of Doradus and Tucana, respectively. They lie about 160 000 light-years away and are companions to the Milky Way.

Figure 5.13. NGC 1365 in the southern-hemisphere constellation Fornax is a good example of a barred spiral galaxy.

Figure 5.14. Galaxy M82 is an irregular galaxy that looks like it is tearing itself apart. The activity is really the result of massive amounts of star birth and star death. The bluish band is the light from hot young stars. The red, filamentary clouds are 10 000-light-year-long jets of material being driven out of the central region by winds from newly forming stars. This superwind helps astronomers put together a picture of stellar evolution and how it shapes the galaxy.

Figure 5.15. Galaxy NGC 3079, where a lumpy bubble of hot gas is rising from a cauldron of glowing matter. (Inset) A close-up view of the central region of the galaxy where a 3500-light-year-long wind tunnel of gas is being driven by the winds of star birth. The action here has taken place over the past million years and as the spurts of star birth settles down, the gas will "rain down" over the galaxy, possibly spurring more episodes of star birth.

Figure 5.16. A look at the Perseus spiral arm of the Milky Way galaxy reveals large-scale evidence of star formation and star death in the interstellar medium of our own galaxy. With information from the Infrared Astronomy Satellite (IRAS), and a radio survey done with a radio telescope array in Canada, astronomers found the unique signatures of hydrogen gas in several of the clouds (colored pink) in the region. Ongoing formation of massive hot stars and the possible passage of a shock wave drives the wispy, unsettled appearance of the area.

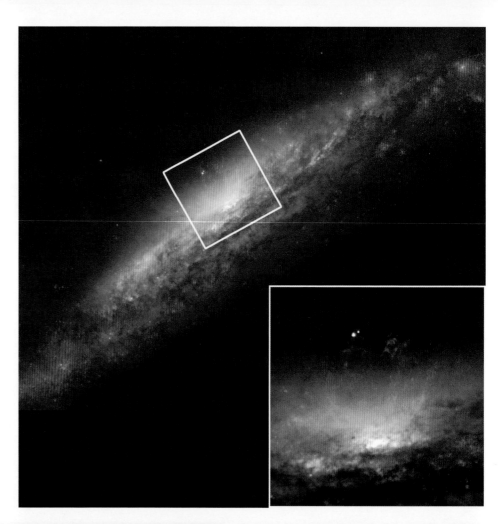

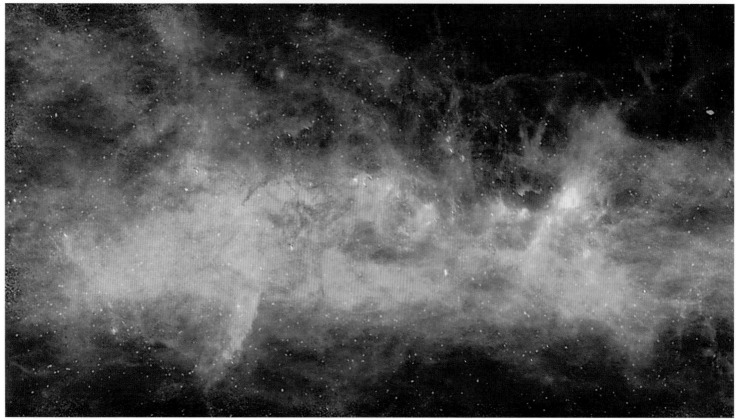

Figure 5.17. NGC 1316, a giant elliptical galaxy in the Fornax cluster of galaxies. Many dark dust clouds and lanes are visible. Some of the star-like objects in the field are globular clusters of stars that belong to the galaxy.

universe, waves of star birth took place over large scales and appear to have been the result of galactic collisions. It also seems now that the number of interacting galaxies is large, and the processes of disruption and merger play some role in the development of spirals.

The central bulges of spiral galaxies assume different sizes and shapes – ranging from extensive, almost globe-like proportions to small and nearly point-like. *Barred spiral* galaxies are a special case. They have long swaths of stars passing through their central regions. These cores generally have older-type stars, although there are nearly always starburst knots arrayed around them.

While the most conspicuous shape of galaxy is the spiral, *elliptical* galaxies are the most numerous in the universe. Usually these have smooth rounded outlines without any spiral shape or structure and can be nearly spherical to almost lenticular. Most of them are populated with millions or billions of very old stars, and don't have too much of the interstellar material needed to birth new stellar generations.

The other entries in our general list of galaxy shapes are the irregular galaxies and they make up about a quarter of all observed galaxies. These can't be classified as either spiral or elliptical. Astronomers have identified two kinds of irregulars – those with lots of star-birth activity (i.e., gas and dust clouds, but no

Galaxies: Tales of stellar cities

spiral arms), and those that are somehow distorted (possibly due to collisions with other galaxies). A few galaxies lie outside the general bins into which astronomers sort them. These odd denizens have unusual structures or exhibit some strong activity and fall into categories called *compact, dwarf, peculiar,* and *active galactic nuclei.*

This multiplicity of shapes led astronomer Edwin Hubble to propose a galactic sorting scheme in the 1920s that at first had possible connotations for galactic evolution. His idea was to classify ellipticals by the extent of their ellipticity – with E0 being nearly circular, E3 being more ovoid, and E7 being almost cigar-shaped. Spirals fell along two not-quite-parallel lines, and were labeled Sa through Sc for increasingly open spirals with central bulges decreasing in size; and SBa through SBc for ever-widening barred spirals. Irregular galaxies didn't fit in either sequence, so they were put off in their own category. When Hubble laid out his classification, the design quite naturally fell into a "tuning fork" shape. Nowadays the diagram has been expanded to include compact, dwarf, peculiar, and active galaxies.

Galactic evolution

Armed with an ever-expanding classification scheme, astronomers have been trying to understand how galaxies came to have their shapes. Certainly the search for an understanding of the evolution of our own galaxy – thought to be a barred spiral with a black hole at its heart – drives a great deal of curiosity about how all galaxies came into being. Because stars evolve, people have always strongly suspected that galaxies must also change over time. However, as we discussed earlier, proving that they do by going out and watching them change is impossible because any galaxy-sized mutations and realignments take millions or billions of years to unfold. As with stars, there are ways around this problem. Just as clusters of newborn stars give us insight into the different kinds of stellar babies that are born, so can clusters of galaxies serve as the samplers of galactic evolution.

The best way is to look for galaxy clusters at different distances – which, because of light travel time, is equivalent to looking back at these structures in earlier times. As observers direct their attention toward more and more of these collections, they have produced some general observations about how galaxies seem to change. First, rotation of galaxies induces density waves that ripple through the stars and interstellar medium like waves on a pond. Second, some galaxies also flatten as they spin, spreading their material out in a wide almost plate-like pattern. In galaxies with active regions, explosive disturbances from supernovae or starburst knots, and core activity, sends huge clouds of gas rippling through near-galactic space. Gravitational attraction between galactic neighbors that merely pass by one another warps and distorts their appearance in a process called tidal disruption. Galaxy mergers and collisions catastrophically alter their structures, sending waves of energy and material plowing through interstellar and intergalactic space. In most cases collisions *and* near misses set off the kind of motion in the interstellar medium that ultimately leads to clouds of gas and dust coalescing to form new stars. As we discussed in chapter 4, the gas and dust clouds that contain the seed material for clouds need a kick-start to get the process going. Shock waves induced by galaxy collisions are as good a way as any to provide the necessary push toward starbirth. As all these mass transformations – star birth, rotation, and interaction – take place, so the appearance of the galaxy changes. Detailed images of distant galaxies are giving us more pieces of the galaxy evolution puzzle, but astronomers don't yet have all the parts to put together the big picture.

Looking out to the galactic deeps

Galaxy studies are being transformed by so-called deep-field surveys that delve into the cosmos searching out the earliest forms of galactic life. One of HST's most significant – and aesthetically beautiful – contributions to the study

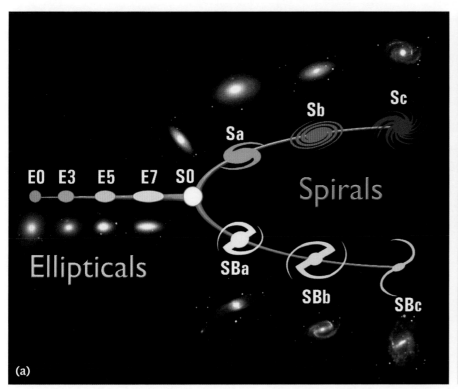

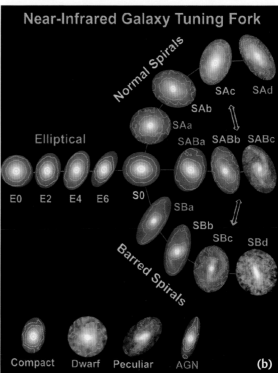

Figure 5.18. **(a)** The Hubble "tuning fork" diagram provides a good visual way to sort galaxies by their shapes. In this version, the galaxies are labeled Sa, Sb, and Sc to indicate spirals with increasingly open spiral structure. These are also referred to as early, intermediate, and late-type galaxies respectively. Spirals with bars are labeled SBa, SBb, SBc to identify the additional component of bar size coupled with the structure of the spiral arms. Elliptical galaxies are identified by their increasing ellipticity with the range E0–E7, with E0 indicating a spherical galaxy. Type S0 originally was a hypothetical transition type. Examples are now known, and the image above S0 shows an edge on view of such a galaxy. **(b)** In recent years, with the help of infrared galaxy surveys, the venerable diagram has undergone a facelift and now looks more like a pitchfork. Elliptical galaxies are still classed on a scale beginning with E0 for ellipticals with circular disks, through E7 in order of increasing elongation. Spirals are labeled as SAa, SAb, SAc, or SAd as the arms open up and the central bulges get smaller (as compared to the size of the galaxy). Barred spirals have their own parallel line and are labeled SBa, SBb, SBc, or SBd. The central tine is a further classification of spirals into more subcategories. Irregular appear in the categories marked compact, dwarf, peculiar, and AGN (active galactic nucleus).

(a)

(b)

Figure 5.19. **(a)** This interacting elliptical–spiral system is undergoing disruption as the two galaxies pass by each other. NGC 5090 (the galaxy on the lower left) appears to be losing material from its outer spiral arm to the nearby elliptical NGC 5091. **(b)** The strongly warped disk of galaxy ESO 510-G13 was created in a collision with another galaxy. The interloper has been gobbled up in the merger. Gravitational forces are reshaping both, but eventually the disturbances will die out. Eventually, this will become a normal-appearing single galaxy. The twisted disk contains dark dust and bright clouds of blue stars whose formation was spurred by the commingling of galactic material, compressing and smashing clouds of gas and dust together.

Figure 5.20. **(a)** The spectacular barred spiral galaxy NGC 6872 is accompanied by its smaller companion called IC 4970 (just above the center). Interactions between these two have pulled long arms of gas and dust from the core of the larger galaxy. The upper arm of NGC 6872 is studded with star-forming regions. There are also many other, fainter and more distant galaxies of many different forms in the field. The bright object to the lower right of the galaxies is a foreground star in the Milky Way. **(b)** A collection of four galaxies, known as Hickson Compact Group 87 (HCG 87), seems to be performing an intricate dance orchestrated by the mutual gravitational forces acting between them. The group's largest galaxy member (HCG 87a) is actually disk-shaped, but tilted so that we see it nearly edge-on. Both 87a and its elliptically shaped nearest neighbor (87b) have active nuclei. A third group member, the nearby spiral galaxy 87c, may be undergoing a burst of active star formation. The gravitational tugs between them is forcing hot gases to flow through the galaxies, accelerating knots of star-birth activity. These three galaxies are so close to each other that gravitational forces disrupt their structure and alter their evolution.

(a)

(b)

145

Figure 5.21. Stephan's Quintet is a well-known grouping of interacting galaxies that lies about 270 million light-years away in the constellation Pegasus. Starburst regions in three of the five are under study in this Hubble Space Telescope view. At least two of the galaxies have been involved in high-speed collisions that ripped stars and gas away from their home galaxies and tossed them into space. But the galactic disaster has also spurred bursts of star formation, resulting in more than 100 star clusters and the creation of several dwarf galaxies.

of galaxy evolution was a project called the Hubble Deep Field. It was a way to "take a census" of typical galaxies in a single small patch of sky in the constellation Ursa Major. That spot was chosen because it was almost empty of nearby stars and the telescope could observe it continuously for 10 days. The field of view of this galactic survey – which you could cover with the tip of your finger held at arm's length – contains nearly 3000 galaxies down to the faint limit of the images. The faintest galaxies are visual magnitude 30, nearly 4 billion times fainter than your eye can see. A second survey of galaxies in the southern part of the sky yielded an even richer lode of young galaxies. Astronomers estimate that we are looking across 12 billion light-years of space, back to a time when the universe was a fraction of its present age. The amazing thing about these deep-field surveys is that if we could look in every direction in the universe with the same intensity that Hubble did (and we didn't have to contend with our own galaxy getting in the way), we would see a "backdrop" of galaxies everywhere. There are millions and millions of them out there, at nearly every epoch of time, bristling with starburst activity, processing matter through stellar nuclear furnaces, and recycling it back into the universe to create the next generation of stars.

The amazing distances and the ages of the objects that HST has seen in its deep-field surveys beg the question, "How do they know how far away these things are?" One way to determine their distances is to take spectra of the galaxy and

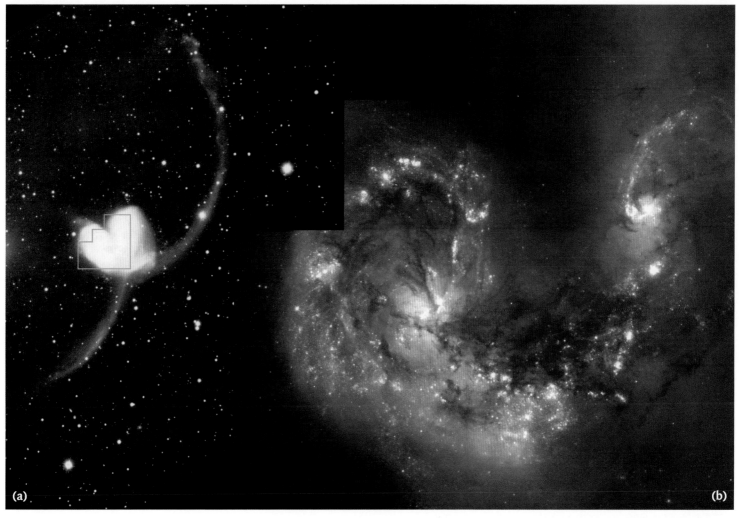

(a)

(b)

Figure 5.22. **(a)** A ground-based telescopic view of the NGC 4038/4039 interacting galaxy pair located 63 million light-years away in the southern constellation Corvus and known informally as the "Antennae." Two long tails of luminous matter, formed by the gravitational tidal forces of their encounter, stretch away from the collision and resemble an insect's antennae. **(b)** The Hubble Space Telescope zeroed in on the merger itself. The respective cores of the twin galaxies are the orange blobs, left and right of image center, crisscrossed by filaments of dark dust. A wide band of dust stretches between the cores of the two galaxies. The sweeping spiral-like patterns are festooned with bright blue star clusters, the result of a firestorm of star-birth activity triggered by the collision.

analyze the wavelengths of light to look for its redshift. As the universe expands, distant objects are moving away from us – and this has an effect on the light we see streaming from them – it literally shifts the light toward the red end of the spectrum. A very distant object will have a very large redshift, and so an astronomer will quote its redshift as a quantity called z. Further analysis of the spectrum will usually give a good estimate of just how far away the object lies.

The knowledge of distances to the galaxies in the Hubble Deep Field also allows us to estimate how much star formation occurred in those earliest cosmic times. The best tracer of star-birth activity is the presence of hot, massive stars, which happen to be extremely bright in ultraviolet light. So, if an astronomer takes

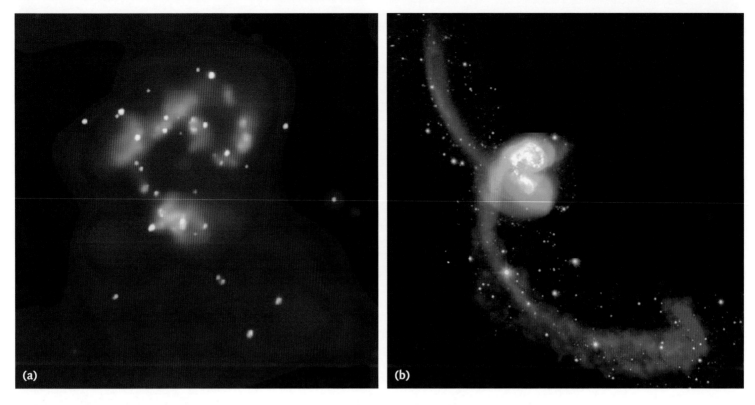

(a)

(b)

Figure 5.23. When astronomers look at the Antennae in x-ray wavelengths they find a large population of extremely bright sources in the same area as the starburst regions that gave rise to the massive young blue stars seen in visible images. These x-ray sources, which emit anywhere from ten to several hundred times more x-ray power than similar sources in the Milky Way, could be massive black holes beaming jets of energy toward Earth. **(a)** This x-ray image shows the central regions of the two galaxies. The dozens of bright, point-like sources are neutron stars or black holes pulling gas off nearby stars. **(b)** In the false-colored image, optical starlight is depicted in green and white, with radio emissions from hydrogen gas coded blue. The radio observations map the distribution of the gas in the colliding galaxies, and help astronomers determine the motions of the pair.

images of a distant galaxy in different colors and sees a bright signal redshifted in one of the images, it's a good bet that the light is from huge numbers of hot, young, massive stars that were were all born about the same time. Images of distant galaxies in different colors can help determine the redshift or z value (which gives the distance) at the same time that it measures the brightness of star formation in the galaxy. If an object is highly redshifted, not all of the original ultraviolet light makes it to our detectors but some does manage to get through. For really distant galaxies however, the ultraviolet light from these stars seems to "drop out" of some images and appear in others. The location of the break depends on the z value (redshift) of the galaxy.

These distant galaxies often do not look like the normal galaxies we are used to seeing. Sometimes they appear to be wispier, or have a shredded look, as if they were still forming. This led astronomers to wonder if there was something different about the actual structures of the galaxies back in their early epochs. Their redshifted ultraviolet light gave astronomers an idea that these actually *are* normal galaxies that look like oddballs because we're only seeing the brightest parts – the star-forming regions. Based on several surveys of nearby galaxies with star-forming regions, astronomers are now looking back at the distant, unusual-looking galaxies to search out the concentrations of newborn stars emitting bright ultraviolet light that is redshifted to the visible range of the spectrum.

The results from the Hubble Deep Field helped astronomers make some very good estimates on the distances to these galaxies and their rates of star formation, and allowed them to come to some general conclusions about how current star-birth rates compare. Many of the galaxies lie about 8 to 10 billion light-years away – and at the time we see them, stars were forming at a rate about 12 times

Figure 5.24. Like archaeologists at a dig on an ancient site, astronomers had to figure out the sequence of events that occurred here 500 million years ago, and determine which nearby galaxy plowed through the heart of the Cartwheel. Two neighboring galaxies could be interlopers, but some astronomers think that the culprit involved in the collision has long since fled the scene of this cosmic traffic accident. In the aftermath of the collision that destroyed the Cartwheel's spiral structure, Hubble Space Telescope resolved knots of star creation in the wake of the collision. The entire outer ring is studded with clusters of hot, young, massive stars created when the collision sent out a ripple of energy that compressed the clouds of gas and dust, causing new stars to coalesce. In other areas supernovae were set off by the dynamics of the collision. The ring contains several billion new stars.

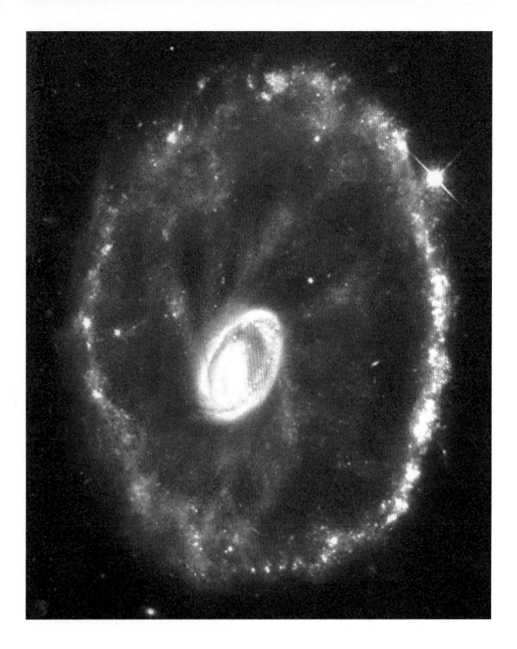

greater than they are today. Now the challenge is to use this information to determine what types of galaxies are responsible for the peak in star formation. Are they spirals? Colliding galaxies? And are their star-birth nurseries located in spiral arms, long tendrils, or arrayed around their cores? Further spectroscopic study should help solve these mysteries.

While it's interesting to look further back in time and farther out in space to see the earliest (and youngest) galaxies, it's only logical to wonder about how they got started. They probably didn't form as a uniform whole (with their present masses) but instead began as sub-galactic clumps. Over time the massive blobs merged to make the galaxies we see today. To test that idea, astronomers used HST to make observations of a patch of sky in the constellation Hercules. What they found supports the "merging" proto-galaxy idea. Eighteen sub-galactic clumps shine from the ultraviolet light of several billion stars that is redshifted into the visible light range. Ten of the sub-galactic clumps have ground-based spectra which

Figure 5.25. When astronomers used Hubble Space Telescope to search for the most distant galaxies, this image was the result. Several hundred never-before-detected galaxies are visible in the Hubble Deep Field image. Along with the classical spiral and elliptical shaped galaxies, there is a bewildering variety of other galaxy shapes and colors scattered across this narrow keyhole view. Some of these galaxies may well have formed less than 1 billion years after the Big Bang.

Figure 5.26. Several years after completing the first Hubble Deep Field exposures, scientists aimed the Hubble Space Telescope at a spot in the constellation Tucana, near the south celestial pole. As with the northern Hubble Deep Field pictures, an astonishing number of different looking galaxies appear in the image. Beautiful pinwheel-shaped disk galaxies with bluish knots dominate the image. But, it also contains peculiar-shaped colliding galaxies. Elliptical galaxies appear as reddish blobs. The colors in the pictures represent the different populations of stars in these distant galaxies. Blue corresponds to young hot stars. Red may indicate older stars, starlight scattered by dust, or very distant starlight that has been stretched to redder wavelengths by the universe's expansion.

Figure 5.27. Look at a galaxy in ultraviolet light and you just might find evidence of past starburst activity. Astronomers studied the galaxy NGC 3310 (which lies 46 million light-years away in the constellation Ursa Major) to map out the existence of super-hot young stars that radiate brightly in the ultraviolet. The galaxy appears to have a good, evenly distributed mix of old and young stars, although the brightest ones trace the outline of the spiral arms. In most galaxies, stars are segregated by age, making classifying the distant ones more difficult. Examples like this one allow astronomers to look for similar structures in the distant universe that are emitting red-shifted ultraviolet light.

Figure 5.28. True-color images of 18 possible galaxy building blocks taken by the Hubble Space Telescope. There are approximately 11 billion light-years away, and they each shine from the light of several billion stars. These sub-galactic clumps could have merged to form the more familiar galactic forms we see today.

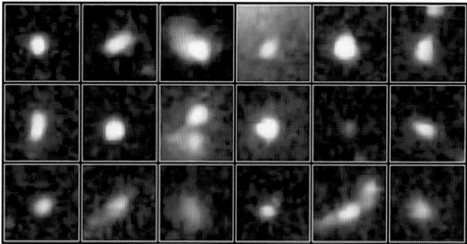

confirm their distance at 11 billion light-years. Each is some 2000 to 3000 light-years across. The next step will be to explore even earlier epochs of galaxy formation and trace the path of proto-galactic evolution from these merging clumps to the sizes and masses now typically found. Since the density in the earlier universe was much higher than now, collisions were likely more frequent, flinging together these massive clumps – the building blocks of galactic life.

Galaxies: Tales of stellar cities

Active galaxies

It is hard to believe when we look at a galaxy in the sky, or an image in a book, that its core may play host to some of the most frenetic activity in the universe. For a long time, galaxies were thought to be quiescent. That changed in 1943, when astronomer Carl Seyfert drew attention to a handful with unusual properties. In visible light, they look like spirals, but they emit strong radio signals and are quite bright in infrared wavelengths. The brightness of the core varies over short periods of time (usually less than a year), which means that the source is quite small. Detailed spectrographic studies of these active regions point to something like a jet of fast-moving material streaming away from the cores.

While the Seyferts are the most active ones, astronomers now recognize huge variations in the activity levels of many galaxies. Our own Milky Way, once thought to be relatively quiet at its center, is actually a strong emitter of radio signals. At the other end of the activity scale, the superluminous quasi-stellar objects – quasars – are almost impossibly bright and energetic. The search is on for a mechanism to explain this range of galactic vigor. Astronomers think they have a good candidate in black holes.

Not long ago, black holes were little more than an interesting theoretical construct. Black holes are formed when a huge amount of matter collapses in on itself so catastrophically that it condenses into a very dense region with an insatiable gravitational pull. This massive pinpoint is called a singularity. The gravitational effect of the singularity is so strong that nothing that passes too close – not even radiation – can escape it.

Speculation about how an object could be so dense that light could not escape from it goes back at least two centuries. In the popular imagination, black holes are fantastic places that somehow transport science-fictional spacecraft across otherwise uncrossable distances in space. It's an interesting concept, but the reality is probably that anything entering a black hole would be crushed so thoroughly that even if it did make it through to some other place, it would be an unrecognizable lump of matter.

Today, astronomers and physicists know that black holes exist. What's more, they invoke them to explain many rare and exotic happenings in galactic nuclei. To understand why a black hole can affect its surroundings, look at its structure. The singularity is surrounded by a surface boundary called an event horizon. It's the theoretical point at which the gravity of the black hole is so strong that nothing can get away. The radius of the event horizon itself is calculated by taking the mass of the black hole (in solar masses) and multiplying it by 3 km. If the Sun were to collapse into a black hole (impossible in reality), the radius of its event horizon would be 3 km. Its diameter would be twice the radius, or 6 km. If a 20 million M_S black hole lay at the heart of the solar system, its event horizon would lie at about the orbit of Mercury.

If you are outside the event horizon, you won't necessarily be sucked into the black hole. Take that fatal step over the event horizon and you're pulled in with no way of escape. Think of the event horizon is the ultimate firewall, hiding whatever there is in the black hole from our sight (and our detectors). Because light cannot

Figure 5.29. **(a)** The elliptical galaxy M87 (which lies some 50 million light-years away), has a giant jet of material streaming from its core – detectable in visible, ultraviolet, x-ray, and radio wavelengths. **(b)** A schematic diagram of a black hole and its environment illustrates how jets like M87's are formed. The accretion disk (red) surrounds the black hole. Its magnetic field lines twist tightly to channel outpouring energetic particles into a narrow jet.

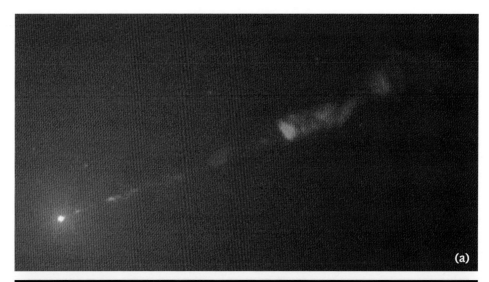

(a)

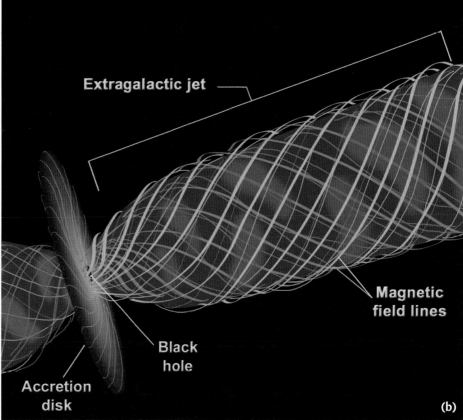

Extragalactic jet

Magnetic field lines

Black hole

Accretion disk

(b)

escape from a black hole, there's no way to know what's happening inside the event horizon. We can only speculate based on what is occurring outside the black hole.

Black holes come in three varieties: mini, stellar, and supermassive. *Mini black holes* are still theoretical and none has been observed. If they exist, they could have formed in the immense pressure, temperature, and turbulence shortly after the Big Bang. They would have relatively tiny masses (comparable to Earth-type mountains) and after billions of years, would simply evaporate away in a flash of energy. *Supergiant stars* at the ends of their lives form stellar black holes. When a star with a core mass equal to or greater than three solar masses exhausts its

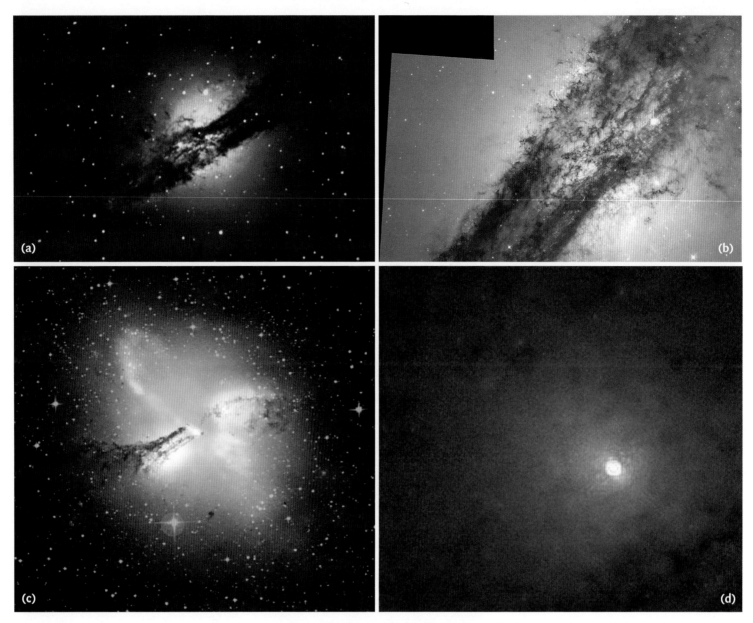

Figure 5.30. Multi-wavelength views of Centaurus A using ground-based optical and radio telescopes, the Chandra X-Ray Observatory, and Hubble Space Telescope, reveal a galaxy in the throes of cannibalizing another one into the black hole at its heart. The resulting chaos has sent waves of energy crashing through the surrounding clouds of gas and dust. In response, they began to coalesce, forming thousands of new stars near the galactic core. In the ground-based **(a)** and Hubble Space Telescope wide-field **(b)** images, the dust lanes are clearly shown as well as star formation sites and clusters of newborn stars, which appear as blue areas. At lower left **(c)** a Chandra X-Ray Observatory image (blue), an optical ground-based image (yellow–orange), a ground-based radio continuum image (green), and a ground-based radio 21-cm image (pink) were all combined to illustrate the result of a frenetic activity at the heart of the galaxy. Multi-million-degree x-ray-emitting gas forms two arcs on the outskirts of the galaxy. The arcs appear to be part of a 25 000-light-year-wide shock front which may be the aftermath of a titanic explosion that occurred about 10 million years ago. At lower right **(d)**, a Hubble Space Telescope infrared image penetrates the galaxy's dark dust lane and shows a twisted disk of hot gas swept up in the black hole's gravitational whirlpool.

Figure 5.31. While jets are the most obvious effect of a supermassive black hole, finding an accretion disk goes a long way toward proving that a black hole exists at the center of a galaxy. This Hubble Space Telescope image shows the central region of the active galaxy NGC 4261. This was the first direct image of a dusty disk. Measurements of its light allowed astronomers to determine the black hole's mass from the disk's rotation rate of about 420 km per second. The V-shaped feature at the center is a strong jet emanating from the center of the disk where the black hole lies.

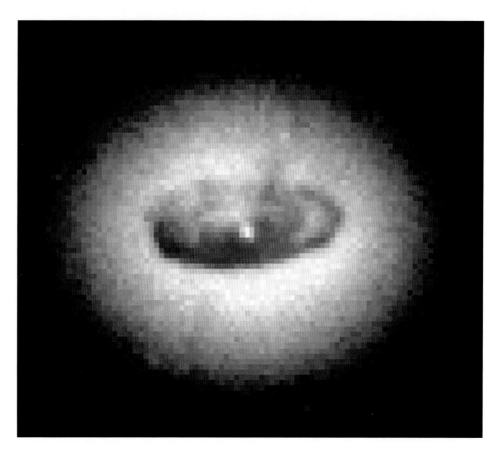

internal energy source it collapses, and produces a supernova. A black hole is formed in the process. *Supermassive black holes* appear to be an essential component of the cores of many galaxies. Conditions in these crowded stellar inner cities seem ideal for the formation of massive stellar black holes, which have merged or gathered up material to form larger and larger black holes of millions or billions of solar masses. Not content to just sit there doing nothing, the supermassive black hole starts sucking up stars, gas, and dust, anything that happens to get too close.

Eventually a huge accretion disk forms around the black hole, feeding material down into the gravitational maw. But it doesn't do this quietly and we can detect the results. There's a tremendous amount of energy released from the region around a galactic black hole. All that infalling material picks up energy as it travels closer to the event horizon. Collisions between objects in the matter stream create heat and light, and a great deal of that light manages to escape. As the accretion disk spins, the material in it is jostled in the gravity well of the black hole and the magnetic field is enhanced. The motion of the disk tugs the magnetic field lines into tightly bound spirals. Charged particles produced by the heating of infalling material fly out along the field lines and are accelerated to near the speed of light, forming a superheated jet of plasma that stretches out well beyond the galaxy. As the mass of the central black hole increases, the jets become more intense, enriching intergalactic space.

So, while astronomers can't detect the black hole directly, they can look for radio waves, strong ultraviolet, x-ray, and gamma ray wavelengths of light that are given off in the superheated environment. Not every emitter of high-energy radiation is necessarily a black hole, though astronomers generally are comfortable calling something a black hole when they see numerous circumstantial effects that imply the existence of one.

What is this circumstantial evidence? NGC 5128, commonly known as Centaurus A, is a likely candidate harboring a black hole at its heart. The core pours out strong radio signals and x-ray emissions, and has supermassive jets streaming out across millions of light-years. All this activity may have begun with the merger of a small spiral galaxy and Centaurus A about 100 million years ago. The merger probably triggered bursts of star formation and violent activity we see in the nucleus of the galaxy. The tremendous energy radiated when a galaxy becomes "active" can have a profound influence on its evolution.

Quasars

We close out our discussion of galaxies with a brief look at quasars. These bright, distant objects were originally detected as radio sources and because they looked star-like, they were first dubbed "quasi-stellar radio sources" – or quasars. They appear to be extremely active galaxies, and for that reason, it's useful to look at what powers these distant beasts. Observationally, quasars are very bright, but can vary in their brightness on fairly short timescales. Surprisingly, a quasar's brightness or luminosity can fluctuate on a timescale of a day or less. This implies that they are small. Large objects have no way to vary synchronously (i.e., all at the same time) across their surfaces. The shortest time by which something's luminosity can vary is the time it takes light to travel across the entire object. Therefore quasars must be quite small – perhaps no more than a few light-days across (not much larger than the solar system) – to vary as fast as they do.

The requirement of extraordinarily large energy production in a small volume together with the existence of jets in quasars points to the conclusion that supermassive black holes are probably the central engines for quasars. A black hole with a mass of 10 billion Suns ($10^{10}\ M_S$) would have an event horizon 400 astronomical units across. Something as bright as a quasar could be this small, if it were a black hole gobbling up a solar mass of material each year.

If this is the power source, the next step is to determine exactly what it is that powers the quasar. How does the energy of matter falling into the black hole get converted into the distinctive quasar spectrum we see – with a range of emissions from x rays to radio waves? Remember that we established the existence of massive black holes by the gravitational effects on their surroundings. Therefore, if we had independent estimates of the mass of the black hole and the accretion mass rate, we could check the luminosity against the observations.

A comparison of nearby quasars with those existing 10 billion years ago implies that the older quasars were some 100 times more luminous than the newer ones. This means that quasars, like galaxies, evolve over time. If we hypothesize higher mass accretion rates to make higher luminosities, then the masses of the central black holes in quasars must also increase with time, at a rate that was higher in the past.

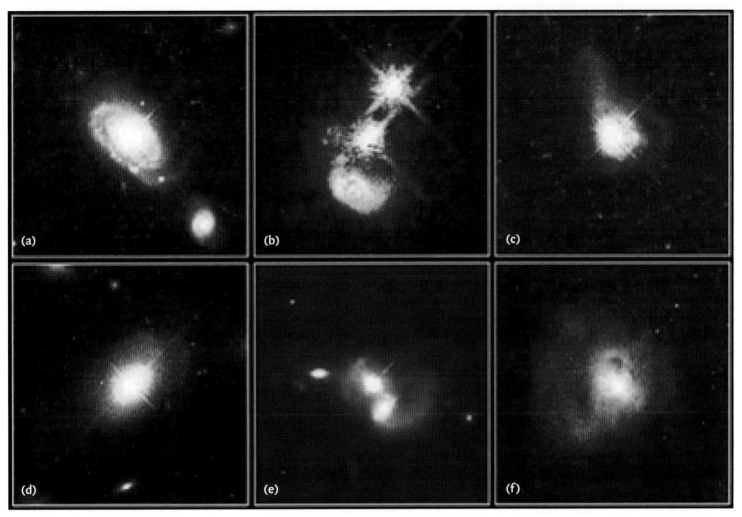

Figure 5.32. Hubble Space Telescope finds quasars in all the right galactic places. **(a)** A quasar shines brightly from the core of a normal spiral galaxy. **(b)** A catastrophic collision between a quasar (central object) and another galaxy (below). A foreground star lies just above the quasar. **(c)** A quasar has captured a tidal tail of dust and gas presumably from a passing galaxy not in the image. **(d)** A quasar at the core of a normal elliptical galaxy, **(e)** a quasar merging with a bright galaxy (just below the quasar), and **(f)** loops of glowing gas that appear to be the result of two galaxies merging.

A fundamental understanding of the light coming from quasars would help answer questions about quasar evolution. Scientists studying quasar spectra have seen broad emission lines in their data. In one interpretation, clouds of gas in orbit around the black hole produce these lines. If this is correct, the speeds of the clouds could be used to determine the mass of the black hole.

Earlier in the chapter we discussed collisions between galaxies and the jump-start they give to star formation. Those same galactic collisions may be one step in the formation of a quasar. But many other quasar images show normal, undisturbed galaxies. Quasars, it seems, are remarkably similar to cats. They behave in mysterious ways, do exactly as they please, and – *so far* – have frustrated attempts to understand their behavior.

6 The once and future universe

God not only plays dice, he throws them in the corner where you can't see them.

Steven Hawking

We had the sky up there, all speckled with stars, and we used to lay on our backs and look up at them, and discuss about whether they were made, or only just happened.

Mark Twain, The Adventures of Huckleberry Finn

Things ain't what they used to be.

Mercer Ellington

At last we come to cosmology. It is a profound topic that encompasses everything about the universe – its origin, evolution, structure, and ultimate fate. There are many philosophical ideas in cosmology that can occupy one's mind while viewing the stars on a dark night. Like Huck Finn, we look up and wonder about the stars – whether "they were made, or only just happened." We might, like all the generations of astronomers who have come before us, seek out the most distant objects, or the most unusual. Or, in a fit of whimsy, we might conjecture whether if we stared long enough into space we'd see the "edge" of the universe. At their heart, these musings are grounded in some realistic questions we can ask (and possibly answer) about the universe. How far away is it all? Where is the beginning? Can we see it? And if we do, what will we find? To answer these questions, we need to answer another one – how old is the universe? Which begs yet another question – will the universe just keep getting bigger, making more stars and galaxies? Or does something else happen to it?

Historically, our perception of the cosmos has expanded every time we figured out a way to magnify our view. Until relatively recently in human affairs, we placed the Earth at the center of the cosmos, and everything else – the Sun, Moon, planets, and stars – revolved around it on nearby spheres. Today our understanding of the universe is less self-centered and more expansive. We know what structures look like in the cosmos, but we're still stuck trying to describe its motion and evolution from our perspective inside of it – much like fish in a stream endeavoring to explain their environment.

Cosmology isn't the most picturesque topic in astronomy – although it certainly uses the pretty pictures and awesome data that astronomy machines churn out. Still, there isn't a specific set of observatory images we can point to and say "That's what cosmology looks like." Understanding the origin and evolution of the universe requires us to draw mental pictures of what we think is happening

Figure 6.1. The Moai of Tongariki seem to be contemplating the splendid isolation of Easter Island in the Pacific Ocean much as cosmologists seek to understand the nature of the cosmos and our place in it.

"out there," using what has been learned about the cosmos from studies of stars and galaxies. Cosmologists (the astronomers who study the evolution of the cosmos) use both direct and indirect methods to study the changes in the universe – and have come up with a vocabulary all their own to express the main ideas of cosmology. In recent years they've settled on a series of key concepts to research and problems to solve as they try to explain the processes that formed and continue to shape the cosmos. At the top of their list are two issues: what are the distances to the most faraway objects in the cosmos, and how much material is there in the universe? Answers to both questions will help astronomers know the age and ultimate fate of the universe. While recent advances in research (coupled with better observations of very distant objects) have brought astronomers closer to a more accurate understanding, there have also been a few new kinks (like the so-called "dark matter" and "dark energy" we discuss below) thrown into the mix just to keep their jobs more interesting.

To be sure, all of our astronomy tools help us peer further back in time, to see ever more distant galaxies and quasars. By performing experiments like the Cosmic Background Explorer (COBE) (which mapped the universe in microwave and infrared wavelengths), the balloon-borne BOOMERanG experiment, and the Wilkinson Microwave Anisotropy Probe (WMAP) – which all map the background of microwave radiation that appears to be emanating from the most distant reaches of space – we have caught the last whispers of the birth of the universe echoing across the cosmos. The International Ultraviolet Explorer opened a window to a universe of frenetic activity in the hearts of distant galaxies, and the Chandra X-ray Observatory is peering deeper into these energetic objects to map their structures. When the Hubble Space Telescope was first conceived, its science mission goals included a myriad of stellar and galactic tasks to fulfill. But, scientists are also using it to answer those big, deep cosmological questions of distance and age.

Cosmological distances and the Hubble constant

We'll start our discussion of cosmology with determination of distances. Humans stand at the foot of a cosmic ladder, anchored on Earth, and stare out to a very distant top rung. We have no way of knowing how tall the ladder is, but we do have the means to climb up (or out) step by step. That's a far cry from our old self-centered place in the cosmos. We now know where we are in the grand scheme of things, and we are learning how far away we are from everything else. Some of our desire to know is driven by simple curiosity: if we measure the physical limits of our universe, we just might be able to understand its origin and evolution.

To solve the great mysteries, astronomers have taken on some fairly esoteric-sounding tasks, like determining a final, unassailable value for the Hubble constant. This much-sought-after number – which in cosmology-speak is referred to as H_0 (and pronounced "H-naught") is a Big Thing – something like the "Holy Grail" of cosmology. If astronomers can determine this figure and accurate distances to faraway objects, they will have a good start toward determining the universe's age, origin, and evolution.

Today's distance determinations to faraway objects are produced through a variety of methods, both direct and indirect (meaning you observe one thing and deduce something about it to help understand another thing), to measure how far it is to objects in the universe. For stars relatively close to us, the usual method is to use a combination of parallaxes (the apparent shift in the position of the star when viewed from Earth over the course of a year) and more esoteric techniques (such as measurements of moving clusters) to determine distances to clusters of stars. If stars in other galaxies and globular clusters can be classified according to the Hertzsprung–Russell diagram (see Figure 4.11, page 100), the main sequence itself can become a tool for determining distances.

In 1924, Edwin Hubble determined the distance to a nearby, naked-eye galaxy he knew as the Andromeda "Nebula." He used a type of star called a Cepheid variable – named after the fourth-brightest star in the constellation Cepheus. Cepheids do something very interesting and useful: they vary their intrinsic brightness over periods of time ranging from 1 to 50 days. Another 20th-century astronomer named Henrietta Leavitt charted the changes for a number of these Cepheid-type stars that lay about the same distance away in the Large Magellanic Cloud. She noticed that the brighter ones seemed to vary over a longer period of time and dimmer ones over shorter timescales. Her observation became the period–luminosity law. If we measure the period of a Cepheid variable (that is, the length of time it takes go from maximum brightness to minimum brightness, and then back again), we can determine its intrinsic brightness. Because the intensity of a star's light fades as a function of how far away it is from us, comparing the intrinsic brightness of the Cepheid with its apparent magnitude gives us a distance. Edwin Hubble applied this method to Cepheids in the Andromeda Galaxy and estimated that it was about 1 million light-years away, thus settling what was then a raging debate about whether or not the Andromeda Nebula was within our own galaxy. (It isn't!) The methods for determining distances and magnitudes have gone through many refinements since that time and today astronomers think that Andromeda is about 2.5 million light-years away.

In today's astronomical parlance, any object whose light you can use to determine distances is a *standard candle*. Think of standard candles like this: imagine that you are standing at one end of a large room looking at a few light sources against the wall at the opposite end of the room. If all of the sources are 100-watt light bulbs, and they are all at the same distance, they should all look equally bright. If some bulbs are closer than others, they will look brighter, but they will not actually *be* intrinsically brighter than the more distant ones. They will always shine as 100-watt light bulbs no matter how far away they are. What astronomers look for are objects that have the same intrinsic brightness – in essence, cosmic 100-watt light bulbs. Some might be a few light-years away and look bright, and others might be a few megaparsecs away and look dim.

Armed with observations of standard candles, astronomers have measured the distances to the Large and Small Magellanic Clouds, the Andromeda Galaxy, and its companion, M32. Now they can extend the distance scale by measuring the brightest stars in other nearby galaxies, and out to more distant clusters. An entire galaxy can be used as a standard candle. Even a specific type of supernova can rival an entire galaxy in brightness and is predictable enough to be used to measure distance.

Another complex method for determining distances relies on analyzing light from galaxies and looking for a telltale redshift in the spectrum. Recall that a spectrum of any object shows characteristic lines for elements either in the object or along the intervening path. As an object speeds away from us, those lines are shifted toward the red end of the spectrum in a phenomenon called a *Doppler shift*. The faster a galaxy is speeding away from us, the greater its redshift. Astronomer Edwin Hubble made the first connection between the redshift of a galaxy (that is, having spectral lines that appear to be shifted to the red or longer wavelengths), how quickly it was moving away from us, and its distance. He came up with a law describing that connection – called, appropriately enough – Hubble's law. For most galaxies, we find that their speed away from us is the product of a number called the Hubble constant (H_0) and their distances. From the cosmological perspective, accurate distances are the key to determining the Hubble constant. Once we have that number in our grasp, we can use it along with a measured redshift of a galaxy to determine its distance accurately.

We mentioned the concept of the cosmic distance ladder earlier. It's easy to think of our place in space at the bottom while the most distant galaxies are at the top. The tools we've described here – determination of distances by using standard

Figure 6.2. A number of methods, all pictured here as the structures making up a lovely cosmological Eiffel tower, are used to determine distances to distant objects. Each method contributes toward finding a reliable value of the Hubble constant (called H_0). Cepheids, the Cepheid variables; TRGB, tip of the red giant branch; SBF, surface brightness function; PNLF, planetary nebula luminosity function; TF, Tully–Fisher relationship; Fornax, Leo I, Coma, and Virgo are clusters of galaxies.

candles like Cepheids, supernovae, and galaxy brightnesses – are just the base of a far more intricate model that takes us out to the limits of the observable universe. Beyond the simple methods, astronomers use a variety of approaches that are more properly the subject of a semester's worth of cosmology coursework. A brief peek at the complete panoply of calculations gives an idea of how complex this work can be.

First, astronomers set the Cepheid variables distance indicators as a broad basis for determining the Hubble constant. Then they look for these stars out to their limit of observability in nearby clusters of galaxies: Virgo, Coma, Fornax, and Leo I. After that, several methods are used to check the accuracy of our Cepheid calculations and to calibrate other methods of distance determination. The first is the Tully–Fisher relationship – which allows astronomers to relate a galaxy's luminosity to the spectral width of its hydrogen emissions at 21 cm (radio wavelengths). The planetary nebula luminosity function (PNLF) lets astronomers look at the brightest planetary nebulae in a galaxy and calibrate their luminosities. Another way is to look at the brightness of elliptical galaxies and measure fluctuations across their surfaces (the surface brightness fluctuation, or SBF method). Some red giant stars – located on the H–R diagram at the tip of the red giant branch – are extremely bright in infrared wavelengths and can also be used to calibrate distances. This is the TRGB method. These indicators, including supernovae (which are very bright but not observed in all galaxies), extend knowledge of distances farther than those for which Cepheids can be used. The goal is to get beyond the nearby clusters (where local gravitational attractions can distort speeds) to the undisturbed "far field flow" at very great distances. Then we will have an accurate, solid value for H_0 (given by dividing an object's velocity rate by its distance). The best current estimate for H_0 is approximately 70 km per second per megaparsec.

So, once we achieve this Holy Grail of cosmological numbers, we can plug in some values, figure out how far the oldest galaxy is, and then we find the age of the universe – right? Unfortunately, no. The Hubble constant is only one of a number of factors that help to create a theoretical construct of the evolution of the universe that long dominated cosmology – something formerly called "the standard model." Think of it as the cosmological blueprint. Cosmologists are the draftspeople refining the numbers that go on the blueprint to come up with a new model. The key numbers they have to deal with are one that tells us when the universe began, and another that describes the expansion of the universe, which is itself influenced by the amount of matter in the cosmos.

The expanding universe and the old, matter-dominated standard model

In the early part of the twentieth century, as techniques to estimate distances to faraway stars and outlying galaxies were developed, Edwin Hubble built on the fundamental discovery that the spectra of galaxies generally showed redshifts. He reasoned that if you interpret this as the Doppler shift (the change of wavelengths of light emitted by an object caused by its motion), it means that the galaxies are all moving away from us. Moreover, when you look at the most distant galaxies, and compare their distances to their redshifts, they appear to be moving away from us the fastest. In other words, the farther away the galaxy, the larger the redshift, and thus, the faster it's receding away.

From this observation, the idea of the expanding universe was established. However, it is important to remember that even though all the galaxies are receding from us, it does not mean that the Earth or the Sun are at the center of the universe. To understand this apparent paradox of thought, consider the popular depiction of an expanding cosmos: a balloon on which we have painted dots to represent galaxies. If we blow up the balloon, its surface expands and *all* the dots recede from each other. There is no central dot. In addition, each dot "sees" every other dot moving away, and those dots farthest away recede fastest. The widely accepted view is that there is no center or preferred location of the universe. Specifically, the location of the Earth or the Sun, or anything else, is not unique in any way.

Along with the idea of an expanding universe and some reliable ways to measure distances, we come face to face with another big question in cosmology: How old is the universe? This is a great question to debate at cocktail parties or on Internet newsgroups. Of course cosmologists do not really ask the question in that way. They are more likely to think in terms of things they can measure, such as the density of the universe, which influences the expansion rate of the universe. So they ask their own types of cosmological questions: How long ago did the expansion of the universe begin? Is the expansion rate the same everywhere in the universe? How does the density of the universe influence its expansion rate? What is the critical density of the universe? How dense does it have to be if it is to expand forever? What density do we need if we want the universe to stop expanding forever? To cosmologists, these are the really big questions.

The answers are complex, and to keep the blueprint relatively simple, cosmologists assign numbers and letters to the key ideas in these questions. We have already talked about H_0, which gives the current expansion rate of the universe in units of kilometers per second per megaparsec. Next, we need to estimate the age of the universe – a variable called "time zero" and labeled as T_0. One way to get to T_0 is to "extrapolate back" to the beginning of time. Take the current rate at which we think the universe is expanding and reverse everything back to a point where it is all back in one place at the same time. You could also think of this as "when time began" – sort of like starting a cosmic stopwatch – and consider some possible numbers to plug into T_0. For the sake of argument, if we use T_0 values somewhere in the range of 10 to 20 billion years, we would get Hubble constants of 100 km per second per megaparsec and 50 km per second per megaparsec, respectively.

Recall now that galaxies twice as far away from us are traveling twice as fast. Thus, when we calculate backward we find that galaxies were all at essentially the same place at the same time – assuming that the expansion rate has been constant. Imagine two cars starting down a highway at the same time with one traveling at 30 mph and the other at 60 mph. At any time, the faster car has traveled twice as far as the slower car. But, calculating backward, we find (of course!) that they both started from the same place at the same time. The method for calculating the time

Open Universe   **Closed Universe**

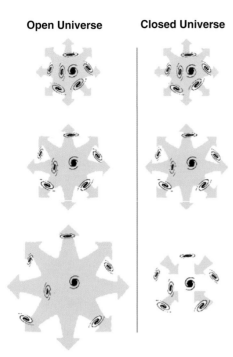

Figure 6.3. What are open and closed universes? An open universe (where Ω_0 is less than 1) expands forever because it does not contain enough mass to slow the universe. A closed universe (where Ω_0 is greater than 1) has enough mass to keep the universe from expanding forever and ultimately it collapses. A universe with exactly the right amount of mass ($\Omega_0 = 1$) is balanced. It expands forever, but more and more slowly as time goes by.

since the Big Bang, T_0, is analogous. If we use $H_0 = 70$ km per second per megaparsec, the age of the universe or time since the Big Bang is approximately 14 billion years.

The higher the value of H_0, the lower the age of the universe; i.e., the faster it expands, and the less time it will have taken to reach its current size. But there is a problem. Simply "plugging in" any value of H_0 without also taking into account the effect of gravity on the rate of expansion gives us a universe that is too old. This is because gravity could have slowed expansion. The big range of ages for T_0 that we mentioned earlier is affected principally by uncertainties in distances and estimates of how much stuff there is in the universe (its density) and the gravitational effect of this "stuff." There is still a lot of work that needs to be done when it comes to determining the true age of the universe. If the density is not accurately determined, then there is no way of setting T_0 correctly.

There are other properties of the expanding universe: Q_0, which denotes deceleration (the rate at which the expansion of the universe slows down), can be used to test the possibility that the exact Hubble relationship may differ for galaxies at different distances. There is also ρ_0 (pronounced "rho naught") – the critical density of the universe, which would be the average density of matter needed to eventually stop the universe's expansion. If we estimate the density of the material there is in the universe today, and divide that number by ρ_0 we get Ω_0. If Ω_0 is greater than 1, the universe will stop expanding and collapse back on itself in something called the "Big Crunch." This state of affairs would give us a "closed" universe. If Ω_0 is equal to 1 then the universe will continue to expand but its rate in the future will be very slow. If Ω_0 is less than 1, the universe will continue to expand, giving us an open universe.

If gravity has slowed things down to the present rate of expansion, then the past rate had to be faster. There are a lot of "ifs" here, and the question of the universe's ultimate fate – indefinite expansion versus collapse and possible renewal – is one of the most fiercely debated in all of cosmology.

Under the "old" standard model, the first instant of the Big Bang is labeled T_0. From there, things happened very quickly. It was followed immediately by an extremely dense, hot universe – an opaque plasma – called the "primordial fireball." Theoretical studies by astrophysicists in 1948 led to the conclusion that a remnant of that fireball existed, probably in the form of a microwave background radiation permeating the entire universe. That signal was discovered in 1965 by Arno A. Penzias and Robert W. Wilson and is called the cosmic microwave background (CMB). This discovery was so important that in 1978 Penzias and Wilson were awarded the Nobel Prize for physics.

The universe as a primordial fireball initially (about 10^{-43} seconds after the Big Bang) consisted of energetic, elementary particles and energetic photons. This is the point where cosmology meets elementary particle physics. At about 10^{-34} to 10^{-30} seconds, the universe was only a few centimeters in size and passed through a stage of very rapid expansion known as "inflation" (when it started with a size less than that of a proton and blew up to about the size of a melon). At around 10^{-6} seconds, the elementary particles (called "quarks") combined to form neutrons and protons. A thousand seconds after the Big Bang, the universe had

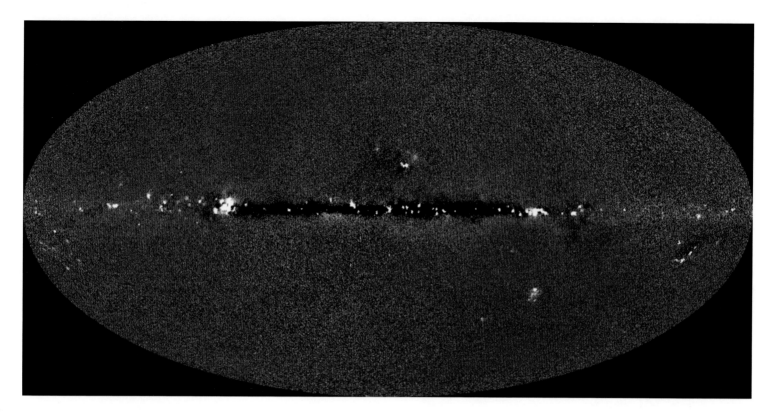

Figure 6.4. After the infrared light from our solar system and galaxy has been removed from measurements of microwave sources, what remains is almost a uniform cosmic infrared background. The line across the center is left behind after galactic light is removed. The Cosmic Background Explorer (COBE) experiment detected a cosmic background of microwave radiation, which is a remnant of the first moments after the Big Bang.

cooled, and the formation of major elements in the primordial fireball was complete. These elements were hydrogen, deuterium (a form of hydrogen), helium, and lithium, and they formed in proportions determined by the density of ordinary matter (also called baryonic matter). The relative amounts of these elements are labeled Ω_B (where B stands for baryonic).

At a point some 300 000 years after the Big Bang the earliest structures in the universe began to form. These regions were places where the earliest matter agglomerations began to coalesce, making it easier to stir up such things as galaxies, and ultimately larger structures like clusters and superclusters of galaxies.

Matter in the universe

Galaxy clusters are important structures both in cosmology and in the study of the evolution of galaxies. They help astronomers get a handle on the distribution of matter in the early universe. Studying their light also helps us infer what other matter might also have been created that is also affecting the expansion of the cosmos.

The biggest known structures are immense filaments and sheets of galactic clusters and superclusters. Undoubtedly, these agglomerations are the remnants of the fine structure seen in the COBE results, which mirror fluctuations in the very early universe that were preserved by its inflationary expansion. Once protogalaxies formed in these structures, the stage was set for the condensation of gas into stars. This was followed by stellar evolution, with its implications for the interstellar medium, the formation of planets, and the development of life.

Figure 6.5. The Virgo cluster of galaxies lies about 55 million light-years away and is the closest large cluster to us. Along with other nearby groups and clusters such as the Local Group of galaxies that contains the Milky Way, Virgo is part of a much larger local supercluster. To the center right of this near-infrared mosaic are the two giant elliptical galaxies, M86 and M84.

Regardless of the state of debate over the origin and evolution of the universe, there is still much work to be done before cosmologists can devise a more modern theory of how it all began and how it all will end. Astronomers have yet to accurately determine H_o, T_o, Q_o, and Ω_o. There is good progress with H_o, but the density parameter (Ω_o) may prove a tougher nut to crack. This is because there are many unknowns about how much matter exists in the universe – and exactly what that matter is. One of the largest unknowns, and an area bristling with current research activity is the existence of a mysterious "stuff" called dark matter. When determining the density of the universe (which affects the expansion), it's what you can't see that can hurt you.

To understand dark matter, it's easier to say what it's *not*: everything you see is not dark matter. Visible matter is called baryonic matter, and is only a small fraction of the makeup of the universe. For decades, astronomers have known that much of the material in the universe is in the form of "dark matter." However, the quantity is unknown.

So, if we can't see it, how do we know it's there? The existence of dark matter is implied because of the gravitational effect it has on galactic motions. Astronomers studying the movement of galaxies expected to see the outer limits of a galaxy rotating more slowly than the inner regions. What they saw instead caused them to surmise that some unseen force or presence was constraining galactic motion. Searches for light-emitting sources fail to find enough mass to account for the motion constraints and the estimated density of the universe. We see the effects of the "stuff," but we don't see the "stuff."

If dark matter coexists with galaxies, is it present in the larger universe? Where do we look for it? Measuring the distribution of galaxies throughout space and comparing it to computer models can help find the answer. In recent years, sky surveys to various depths in the universe have begun to take a census of galaxies, making the comparison more accurate. The 2-Degree Field Galactic Redshift Survey has observed nearly a quarter of a million galaxy redshifts and used the information to come up with a more accurate computer simulation of the universe. It would appear that dark matter with the immense gravity it wields pulls the normal baryonic matter along with it. In other words, where there is light matter, there is an unseen mass of dark matter making its influence felt.

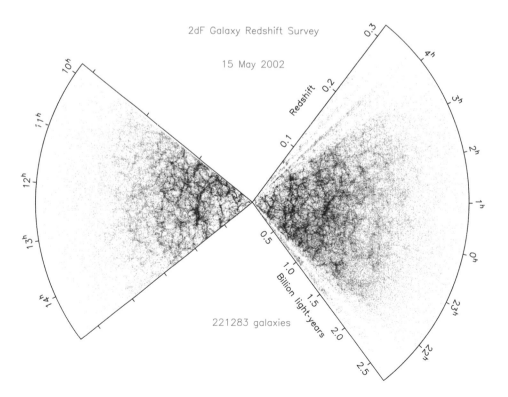

Figure 6.6. These blue cones represent a three-dimensional distance map of 221 283 galaxies. The distances were derived from redshifts taken by the 2-Degree Field Galactic Redshift Survey out to a distance of 2.5 billion light-years. The galaxies seem to be distributed through space in waves, and these patterns imply that there is more than baryonic and dark matter rearranging these galaxy distributions. One possible candidate is dark energy, which hasn't yet been directly detected.

Since we cannot directly observe dark matter, we have come up with many possibilities for what it might be, and some of the names are quite imaginative: Massive Compact Halo Objects or MACHOs – very faint stars or Jupiter-like objects – which could be arranged in an essentially spherical cloud around galaxies. It is also possible that black holes are much more widespread than we suspect. There are possibilities of exotic particles called Weakly Interacting Massive Particles or WIMPs – that would theoretically have the mass to contribute to the dark-matter inventory. They haven't been discovered yet, but particle physicists think they exist. Axions are another possibility – but they too are still in the theoretical stage. These latter two possibilities, if they're really out there, could be remnants of the very dense, hot early universe that existed shortly after the Big Bang.

Just a few years ago, cosmology appeared to be a relatively simple subject. As we've discussed, the two most important numbers we needed to determine were the Hubble constant, H_o, and the density of matter in the universe, ρ_o. Determining H_o to better than 10 percent was one of the "key projects" for the HST. After years of research astronomers determined the current value to be close to 70 km per second per megaparsec.

Now, we need to examine the assumption that the speed of expansion has been constant. We know that matter slows the expansion of the universe due to its gravitational pull. Thus, we have to figure out how much matter there is. As we discussed earlier, if there's not enough matter, expansion would go on forever. If there's too much matter, the universe collapses back on itself. If there's just the right amount of matter, we need to ask if it's slowing down the universe over time.

There are some ways to measure matter to help us determine how old the universe is and what will happen to it. For example, the current ratio of deuterium to hydrogen (D/H) in the interstellar medium is a sensitive indicator of the amount of ordinary matter. Analysis of light from stars that has passed through the interstellar medium tells us that the amount of baryonic matter is small. In terms of the critical density the amount of ordinary matter is roughly 5 percent of the total mass of the universe. If baryonic matter were the only kind of matter in the universe, the density would be well below the critical density and the universe would expand forever.

Moreover, studies of clusters of galaxies find that dark matter is needed to hold them together. This means that the dark matter must be "cold" in order for it to concentrate in these clusters. However, a recent discovery of diffuse ordinary (baryonic) matter was made when astronomers looked at dark absorption lines in ultraviolet and x-ray spectra of distant the light sources. These so-called shadowing effects were caused by something described as "intergalactic fog" that seems to exist in the Local Group of galaxies. So, it seems that at least part of the mass needed to hold these clusters together may have been found.

The exotic alternative, called hot dark matter, would not be concentrated. In our bookkeeping of energy and matter in the universe, 30 percent must be in the form of cold dark matter. If it is, then the expansion is hardly slowed by gravity and T_o remains close to 14 billion years. And the expansion would continue forever.

Figure 6.7. If the energy and matter in the universe are graphed on a pie diagram, the current proportions for the universe look like this. Instead of assuming that the universe is solely made up of stuff we can see, most of the universe is in the form of "dark energy." Ordinary, non-luminous (baryonic) matter constitutes a very small fraction of the total and even "dark matter" is a relatively small fraction. The approximate values are two-thirds dark energy (which has also been called a repulsive force) and one-third total matter.

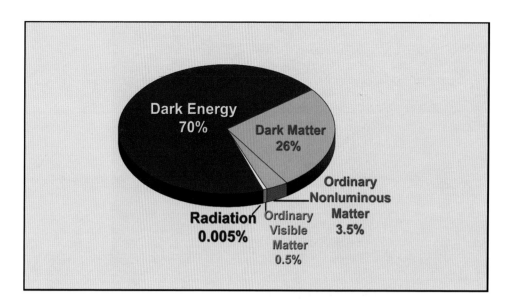

The expanding universe and the new energy-dominated standard model

While the scenario we've been discussing would have been a reasonable summary of evolution of the early universe in about 1997, there were problems that led to some rethinking among cosmologists. Some observational evidence and theoretical predictions argued that Ω_0 was equal to 1. If this were the case, then a major chunk of matter of the universe was missing from our inventory! Everything changed in 1998 when studies of supernovae indicated that the universe was not slowing down, but was, in fact, speeding up. Detailed results from studies of the cosmic microwave background radiation confirmed that Ω_0 was very close to 1. In other words, the universe has a flat geometry. Since we know that the expansion of the universe is currently accelerating, this helped trigger a major change in point of view by providing a missing piece of the puzzle.

At this point a very logical question from the reader should be "How on earth can we know these things and construct a model of the universe with any confidence at all?" The answer is simple in principle and complex in detail. Events and processes even long ago leave traces in the universe. If these traces can be measured and understood, the early history and current status of the universe can be inferred. For example, the cosmic microwave background radiation is a remnant of the universe when it was approximately 300 000 to 400 000 years old. This radiation from a distant time comes to us redshifted and has a blackbody temperature of 2.7 degrees.

Now, let's fill in some details of the logical framework that leads to the summary pie-chart in Figure 6.7. We start with supernovae and the accelerating expansion. Type Ia supernovae are among the brightest objects in the universe and they appear to be "standard candles" (as shown in Figure 6.8). Basically, a star in the white dwarf stage cannot have a mass in excess of 1.4 M_S. If material from a companion star pushes the mass over this limit, the star collapses and triggers a gigantic thermonuclear explosion. No remnant is left behind to decay in brightness. The fact that essentially the same amount of material is involved produces the "standard candle." If they are found in galaxies with distances determined from

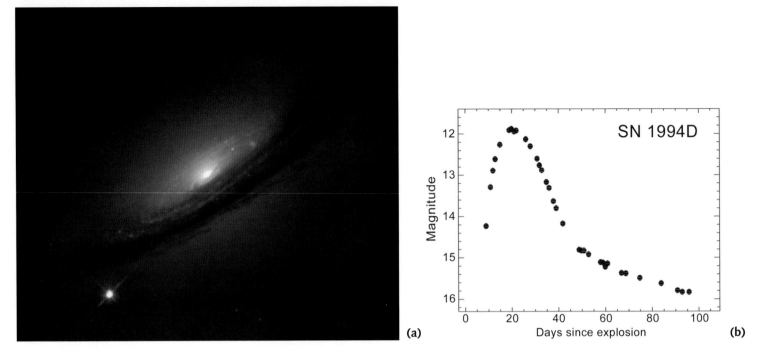

(a)   (b)

Figure 6.8. **(a)** Supernova 1994D is visible as the bright object in the outskirts of the galaxy NGC 4526. This is a Type Ia supernova – an explosion that occurs when material from a nearby companion falls onto the surface of a white dwarf. They are found in many galaxies and serve as standard candles – distance indicators that allow determinations of the universe's expansion rate and its geometry. **(b)** By measuring the magnitude and shape of the light curve of SN 1994D as it brightened and faded we can determine the luminosity of the event and the distance to it.

other methods (e.g., from Cepheids or other methods shown in the H_0 Eiffel diagram), small differences can be calibrated based on the light curves. This comparison is not simple and has been carried out by two different groups who have found that the expansion rate is currently accelerating.

What could be causing the rate of expansion to increase? Currently, the best answer is a sort of cosmic repelling force called "dark energy." Years ago, Einstein had derived a static, but unstable, model of the universe. He introduced the cosmological constant, denoted by the Greek letter lambda – Λ, to overcome the difficulties, but then discarded it. He later called the cosmological constant his greatest blunder. The current best (but not perfect) model is a combination of Λ (for cosmic repulsion) and the cold dark matter discussed above.

This general picture can be checked. The relative effects of the cosmological constant and gravity change as the universe expands. In the early universe, the same amount of matter was contained in a small volume and was actually able to overcome the accelerating effects of the cosmological constant. Thus, the expansion was initially slowed. But when the universe had expanded to a large size, the small but steady effect of Λ took over and the expansion accelerated. Testing this idea requires Type Ia supernovae that are farther away and thus farther back in time. An example has been found (see Figure 6.9). In contrast to the supernovae in the accelerating regime, supernovae in the decelerating model are brighter than expected. Detailed calculations using these distant candles still confirm an age of the universe of about 14 billion years.

Figure 6.9. Supernova 1997ff occurred back at a time when the expansion of the universe was slowing down, shortly after the Big Bang. Later the expansion began to speed up, and this supernova provides additional evidence for the current view of the expanding universe. In this image, a portion of the Hubble Deep Field (top) shows the region containing Supernova 1978ff (inset). The inset region (bottom left) shows the supernova's host galaxy, marked with an arrow. The supernova itself appears in the difference image (bottom right). The mottled red background in the difference image is an artifact of image processing.

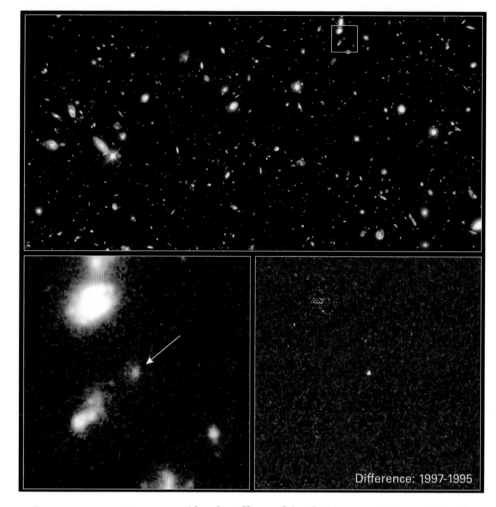

Because we must now consider the effects of the dark energy (Λ) as well as the matter in the universe, Ω_0 as discussed earlier must be redefined to include both dark energy (Ω_Λ) and matter (Ω_{mater}). Thus, $(\Omega_\Lambda) + (\Omega_{matter}) = \Omega_0$ and the value of the new Ω_0 determines the geometry of the universe as illustrated in Fig. 6.10. Specifically, we have a flat geometry for the universe with $(\Omega_\Lambda) + (\Omega_{matter} = \Omega_0) = 1$.

The cosmic microwave background radiation is a gold mine of cosmological information. At the time it was first radiated as light, some 300 000 to 400 000 years after the Big Bang, the cosmos was permeated with sound waves, which produced some lumpiness. At that time, the universe had cooled enough to allow protons and electrons to come together to form neutral elements, which no longer blocked propagation of light. Essentially, all of space became largely transparent, and it is the visible light freed at that time that we now see redshifted as microwaves forming the cosmic background. Cosmologists have taken many measurements of the cosmic microwave background at different resolutions and those measurements argue for a universe that had gas spread in an almost smooth pattern. However, it wasn't perfect. There were irregularities that pre-dated the structures that are today's intricate networks of galaxies. The sum of those observations (made with the microwave-sensitive Cosmic Background Imager in Chile), show bumps and curves that reflect the "lumpiness" of the early universe as reflected in the microwave background.

Figure 6.10. To understand the effect that Ω_0 has on the structure of the universe, think of the cosmos as a series of geometrical shapes, depending on the value of Ω_0. Our ordinary conception of space looks flat, like the $\Omega_0 = 1$ scenario shown here. If the universe is closed, it is finite – like the sphere in the $\Omega_0 > 1$ scenario. An open universe (where $\Omega_0 < 1$), takes on a curious-looking saddle shape. Another way to think about the geometry is in terms of parallel lines. In a flat geometry, lines that are locally parallel remain parallel as they are projected over distances, like the rails on a train track. They remain the same distance apart. But, for a closed universe, they intersect and, for an open universe, they diverge. The red triangles in the figures illustrate different geometrics. For $\Omega > 1$ the interior angles of a triangle sum to more than 180 degrees. For $\Omega < 1$ the interior angles add up to less than 180 degrees. For $\Omega = 1$, the interior angles equal 180 degrees.

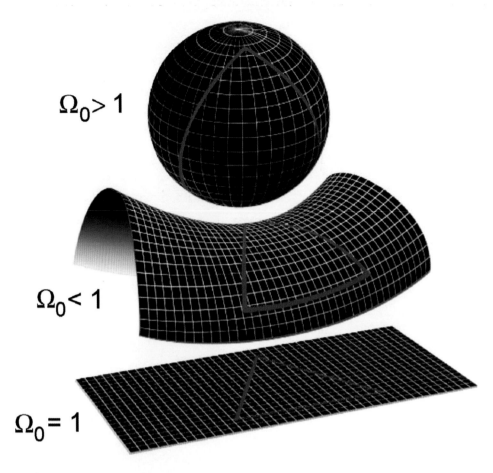

After examining all these methods, the story of the universe now goes something like this: it all began with the Big Bang and shortly thereafter "inflation" greatly increased its size. Inflation magnified tiny fluctuations into the density concentrations that eventually became galaxies, and ensured that Ω_0 was very close to 1. Then, nucleosynthesis (involving normal matter) produced the light elements in the universe. When the density of baryons dropped enough, matter and light decoupled – they were no longer one. The radiation that became the cosmic microwave background radiation was emitted, sending information forward in time and across the light-years about the structure of the early universe. Initially, the Hubble expansion slowed some because of gravity, but after a while the cosmological constant dominated and the expansion accelerated. The universe will probably expand forever, but the ultimate fate cannot currently be predicted with certainty.

Progress in understanding our universe continues at a rapid rate. The Wilkinson Microwave Anisotropy Probe has obtained the first detailed full-sky map of the oldest light in the universe! The microwave light it captured originated some 380 000 years after the Big Bang. Cosmological parameters have been determined from this light by fitting a model to the observations. The age (T_0) is 13.7 billion years. The value of Ω_0 is close to 1 and supports the idea that inflation occurred. Of course the universe has a flat geometry, and the composition derived from this data makes it 73 percent dark energy, 22 percent dark matter, and 4.4 percent ordinary (baryonic) matter. These extraordinary results are consistent with our current understanding, but appear to provide information at unprecedented

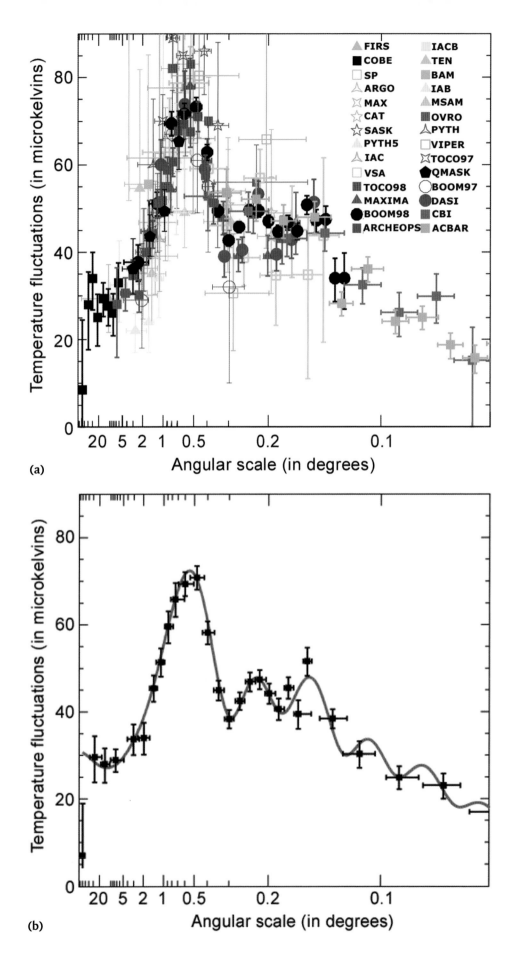

Figure 6.11. (a, b) Combined measurements of the cosmic background radiation shed more light on the early universe. In the first plot **(a)**, many measurements of the strength of the fluctuations are plotted versus the angular size of the fluctuations. In plot **(b)** the various observational results have been weighted and averaged. The red line represents the fluctuations predicted by models of the universe with a mix of ordinary matter, exotic matter, and dark energy. (The perpendicular crossbars are error bars and represent the uncertainty in the measurements.)

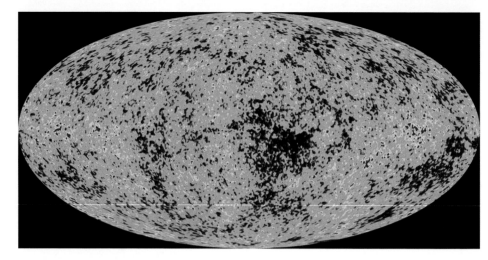

Figure 6.12. The Wilkinson Microwave Anisotropy Probe (WMAP) image of the microwave sky provides a "baby picture" of the universe. The oval shape is a projection of the whole sky. The red colors indicate warmer regions and blue indicates cooler regions of the sky. WMAP can record temperature fluctuations in the millionths of degrees. These fluctuations are the seeds that generated the cosmic structure (galaxy clusters, etc.) that we see today. Interpretation of these temperature changes using a cosmological model provides accurate values for the parameters cosmologists use to describe our universe.

accuracy. The probe is expected to continue providing improved measurements of these early epochs of the universe.

Cosmology is now a branch of astronomy based on many areas of accurate observations. But, there are also worrisome aspects. In our recipe for the universe in Figure 6.7, two of the largest items are dark matter and dark energy – things we have never seen whose presence is inferred only by measuring their side effects. Also, while there are strong indications that inflation took place near the beginning, it's not yet clear what caused it to happen.

Gravitational lenses

Einstein's theory of general relativity is the basis for calculating the evolution of a Big Bang universe. Interestingly, the theory also brings up some intriguing effects that help us study the universe. According to general relativity, the space (strictly speaking, the "space–time") near a massive object is distorted by that object. This produces the same basic effect as a gravitational pull, and orbits can be produced.

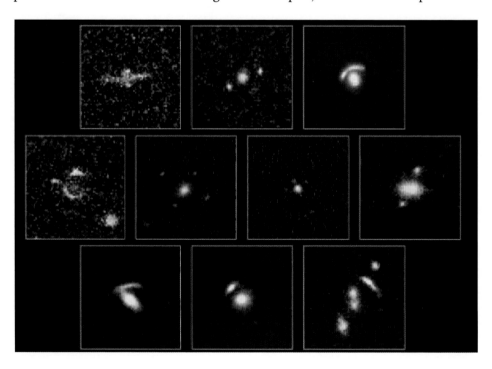

Figure 6.13. The Hubble Space Telescope peered at a number of gravitational lenses and cataloged them into a kind of "top ten" list of shapes. The appearance of a gravitational lens depends largely on our viewing angle, plus the size of the lensing object.

(a)

Figure 6.14. (a) The rich galaxy cluster Abell 2218 serves as a massive gravitational lens. The cluster is so massive and compact that light rays from a very distant population of galaxies are bent to form the arcs of light. Those galaxies existed when the universe was just one-quarter of its present age – about 3–4 billion years old. Since we see these galaxies at a very young age, the arcs can be used to look at star-forming regions in the distant galaxies. (b) As it did for Abell 2218, the Hubble Space Telescope peered straight through the center of another massive galaxy cluster called Abell 1689, considered to be the most powerful gravitational lens found to date. For this observation, Hubble had to gaze across 2.2 billion light-years of space for more than 13 hours. The gravity of the cluster's trillion stars – plus associated dark matter – acts as a 2-million-light-year-wide "lens" that bends and magnifies the light of very distant galaxies, distorting their shapes and creating multiple images of individual galaxies.

(b)

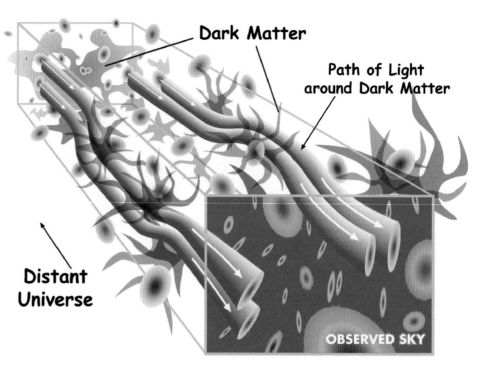

Dark Matter

Path of Light around Dark Matter

Distant Universe

OBSERVED SKY

Figure 6.15. Dark matter along the line of sight to distant galaxies distorts the observed image. Light rays from distant galaxies travel through a universe filled with clusters of dark mass. The figure shows a schematic of so-called "weak" gravitational lensing (where the lens isn't strong enough to form multiple images and arcs, but instead distorts the image). Every bend in the path of a bundle of light from a distant galaxy stretches its apparent image. The orientation of the resulting elliptical images of galaxies contains information on the size and mass of the gravitational lenses the light encounters as it travels. Here, light bundles from two distant galaxies which are projected closely together on the sky follow similar paths and undergo similar gravitational deflections by intervening dark matter concentrations. The larger the mass in the gravitational deflectors, the larger the ellipticity of the faint galaxies appears to be. Correlating these elliptical distortions in the images of faint distant galaxies allows cosmologists to estimate the amount of dark matter doing the distorting and make some better-educated guesses about the amount of dark matter in the universe.

Figure 6.16. This admittedly complex graph summarizes the many methods of estimating the amount of dark energy (Ω_Λ) and the density of matter (Ω_{matter}) in the cosmos and the implications for the future of the universe. WL, weak lensing; CMB, cosmic microwave background; SN, supernovae; QSO, quasi-stellar objects; BBN, Big Bang nucleosynthesis in which the earliest elements were made. The "lower limit" line plotted at approximately $\Omega_{matter} \sim 0.2$ does not have a corresponding upper limit because non-baryonic matter not concentrated in clusters has not been accurately measured. The arrow marks the area where all the estimates converge.

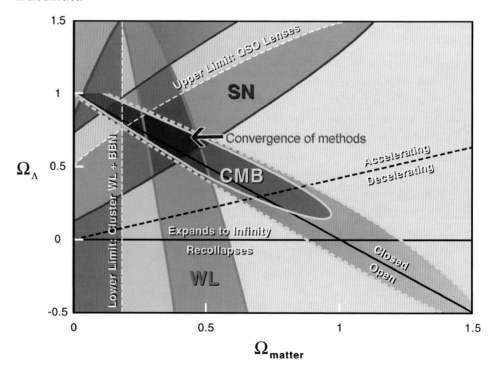

Figure 6.17. If we could present a "snapshot" of cosmology, this grand summary schematic of the universe might be a good candidate. It begins with the Big Bang. After inflation, the universe contained energetic photons and a "quark soup" consisting of electrons, neutrinos, and quarks (the building blocks of protons and neutrons). Nucleosynthesis marks the era in which hydrogen, helium, and a little lithium were synthesized. When decoupling occurred, matter and radiation were separated. This released the early radiation that we see as the faint whisper of the cosmic background, which forms the backdrop for our modern universe.

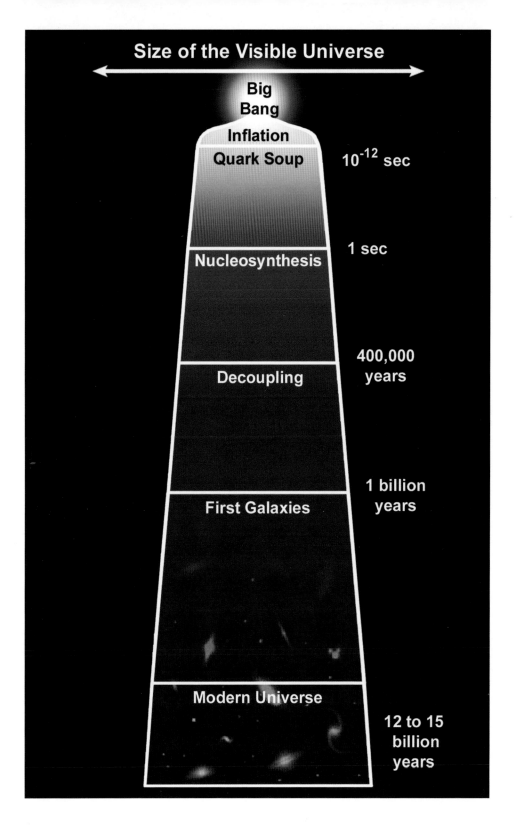

In fact, for planets, general relativity and Newtonian mechanics give only slightly different results.

This distortion of space has another consequence. The light shining from an object directly behind another, more massive, object can have its direction changed so that the object itself appears to the observer to be shifted from its real position.

The once and future universe

An extended object can have its image changed; this phenomenon is called "gravitational lensing." In our brief discussion of the geometry of the universe, we mentioned how geometry can influence or change the appearance of objects. Thus, we can probe the properties of the universe by studying lensing. Theoretically, any body with mass can lens any other body. Practically speaking, however, detecting the lensing on anything smaller than stellar or galactic scales is next to impossible.

The discovery of different gravitational lenses is exciting and studying the phenomenon offers the potential for very important cosmological results – particularly in the search for dark matter. Light traveling through the universe can have its path altered by dark matter just as a baryonic galaxy cluster might distort it. When the distortion is slight, the effect is called weak lensing. Studies of strong lensing in quasars and weak lensing effects elsewhere can also provide important constraints on cosmological models.

What does it all mean?

For now cosmologists have some good estimates (based on all their methods) for many of the parameters needed to calculate the fate of the universe and complete the cosmic blueprint. Those estimates for Ω_Λ and Ω_{matter} are plotted in Figure 6.16. The arrow marks an area where all the different pieces of the puzzle converge. It is consistent with a value of 0.7 for Ω_Λ and Ω_{matter} is 0.3. These numbers produce a flat universe ($\Omega_0 = 1$). The ordinary (baryonic) matter contributes a very small piece of the amount of the Ω_{matter} budget (0.05). What all of this means is that the universe will probably expand forever, but until all the uncertainties in our current understanding are solved, that statement is probably very uncertain.

It would be nice if the grand summary of our understanding of the universe were as easy to understand as a pretty HST picture full of galaxies and lensing. Cosmologists tend to picture these things as graphical models, but in simple terms, the methods they use to estimate the mass of the universe, its expansion rate, and the so-called cosmological distances are all tools that they continue to debate and refine as they get more and better data from their instruments. Each new investigation and discovery gives better insight into the early universe. While we may never be able to peer beyond the curtain hiding the Big Bang from us – into the opaque early universe – cosmologists are working on expanding our view back to the earliest epochs after the Big Bang – to the first structures of the cosmos that ultimately became today's universe.

No one regards what is before his feet; we all gaze at the stars.

Quintus Ennius

Many discoveries are reserved for ages still to come . . . Our universe is a sorry little affair unless it has in it something for every age to investigate . . .

Seneca

7 Stargazing: The next generation

We have just made our way through six chapters illustrating the wonders of the universe. Based on the tremendous range of cosmic objects we've covered here, it might seem to a casual reader that astronomers have discovered all there is to find out about the cosmos. Nothing could be further from the truth! As the first-century AD philosopher Seneca mused in his writings, there are many discoveries waiting to be made.

Even if astronomers *had* found everything in the universe, just cataloging the sheer immensity of what exists is enough to keep researchers busy for countless generations. The universe is far from being "explored." As the late Carl Sagan noted in his *Cosmos* series, humans have waded out perhaps only ankle-deep from the shoreline of the cosmic ocean. There are vast and ever-changing waters farther out awaiting our attention. They are where our greatest opportunities for exploration remain.

Today's dreams for future astronomy observatories will be reality for our children and grandchildren. The situation is roughly analogous to the early decades of the twentieth century, when scientists and engineers had only an inkling of what instruments might be used in the future to explore the cosmos. In 1923, German rocket scientist Hermann Oberth wrote up *his* idea for a space station and an orbital telescope. In his book *Die Rakete zu den Planeträumen* (*The Rocket into Planetary Space*), Oberth imagined a telescope attached to a station in geosynchronous orbit (an orbit that matches Earth's sidereal rotation rate) and discussed such problems as station jitter on observations of dim, distant objects. To solve that problem, he came up with some rather ingenious ideas for anchoring a space-based telescope to a small asteroid. Oberth even suggested that engineers figure out some way to tow the asteroid to Earth orbit so that the telescope's crew would not be bored while living on a station orbiting between Mars and Jupiter!

Figure 7.1. The future of astronomy begins, as it always has, in the minds and hearts of children. Today's explorers were inspired by their parents, by a space race, a television program, a planetarium visit, or simply by stepping out under the starry skies. Tomorrow's astronomers are hot on the heels of their parents and grandparents. Ingrid Braul (left) and Harveen Dhaliwal pose before a huge image of the Trifid Nebula taken by the Gemini Observatory in Hawaii, and projected onto the wall of the H.R. MacMillan Space Centre Star Theatre planetarium in Vancouver, British Columbia. Braul received the Trifid image for her award-winning essay suggesting why she would like the Gemini North telescope to take a closer look at the nebula. Dhaliwal was presented with a Gemini image of Pluto for her essay about the ninth planet.

Oberth's writings attracted a great deal of attention and criticism at a time when any idea of traveling beyond the atmosphere was best left to science-fiction writers. Yet, his ideas seem completely obvious to us *now*, because many decades after he wrote them, astronomers have turned our planet and our solar system into a huge roving eye – capable of seeing out to the limits of the observable universe. They've done this with a collection of orbiting observatories and planetary missions. Not to be outdone, ground-based installations are producing images and data of a quality that is rapidly approaching – and in some cases – rivaling observations from space-based observing systems.

One wonders which of today's outlandish ideas for observatories will be tomorrow's obvious technology. The astronomical drawing-boards are filled with draft plans for an amazing array of new space probes and ground-based arrays – some of them every bit as challenging for us to fathom as Oberth's asteroid-based telescope was to the people of his time. Against a backdrop of technological change, prognostication about future observing platforms is never easy. Those of us who use computers in our work are constantly reminded of the day-to-day changes that almost instantly render any new software releases and hardware changes obsolete. The same is true of the advanced technologies that resulted in adaptive optics, more efficient detectors, and rapidly increasing storage capabilities that characterize the changes in astronomical research. Still, the urge to peer into the future of astronomy is at least as irresistible as peering into the deeps of space (and time). So, to round out our armchair tour of the cosmos, we'll end with a look at a few of the many missions and observatories proposed for the next 10 years and beyond.

Continuing the quest for Mars

The past few decades have seen some amazing feats of *in situ* planetary exploration, starting with the first steps taken on the Moon in 1969. The generation of children born and raised in the last half of the twentieth century watched as their parents and grandparents opened the way to Mercury, Venus, the Moon, Mars, Jupiter, and the outer solar system with a series of planetary missions. These days it's their turn to plan and send the missions that will thrill *their* children and pave the way for their grandchildren to set foot on other worlds.

The planet Mars will continue to receive its share of attention. While some Mars enthusiasts would like to see people standing on Mars in the next few years, the reality is that it may be more likely the second or even third decade of the twenty-first century before humanity takes that "second giant step" to the planets. Until that time, the remote exploration of Mars will continue. From Earth (and HST in Earth orbit) the planet is a favorite for motivated observers (both professional and amateur) who continue to chart the changes in its brightness, the annual cycle of its polar caps, and the periodic dust storms that sweep across its surface.

A suite of mapping and roving missions over the next few years will follow up the successes of recent *in situ* missions like Pathfinder, Mars Global Surveyor, and Mars Odyssey. The most recently launched missions are the Mars Express and two Mars Exploration Rovers. Plans beyond these are still uncertain, but a Mars Reconnaisance orbiter is in the works, and a sample return mission in the second decade is definitely on the drawing-boards.

Mars Express is a 2003 mission using an orbiter and a lander to explore the Red Planet. It was planned by the European Space Agency, the Italian space agency, with cooperation from NASA and the British Beagle 2 project. The orbiter will concentrate on the atmosphere and surface of Mars from polar orbit, searching for sub-surface water and studying the planet's structure and geology. The Beagle 2 lander is an astrobiology mission. Upon settling on the surface of Mars it will perform a series of geochemical and biological tests aimed at determining whether life exists (or has ever existed) on the Red Planet.

In 2004, twin Mars Exploration Rovers will land in two different places on the planet and wander about 100 meters per day across the surface. If all goes well, the rovers will do their scientific explorations for at least 90 Mars days. Their primary goal is to test for evidence of liquid water on Mars in the past, but their first accomplishment will be to take 360-degree color and infrared panoramas, much as the Pathfinder mission did upon its landing in 1997.

The images and spectra returned by the rovers will allow scientists to command the vehicles to explore interesting rock and soil targets and study their compositions in much finer detail. As the missions progress, the rovers will

Figure 7.2. The Mars Express mission will employ a mapper and land the Beagle 2 station to explore the Red Planet beginning in December 2003.

Figure 7.3. One of two Mars Exploration Rovers that are planned for 2004 missions.

wander far afield from their landing pads, covering more ground in one day than the 1997 Sojourner rover did during its entire 77-day lifetime on the planet.

The 2005 Mars Reconnaissance orbiter will be a high-resolution mapping mission orbiting the planet. As with every other Mars mission, the search for water is an important priority, but the planned science objectives also include sweeping the surface for suitable landing sites. The Reconnaissance orbiter's planned instrument package will include a visible stereo-imaging camera to give a highly detailed three-dimensional view of the Martian surface. Alongside the camera will

be a visible and near-infrared spectrometer that will study the planet's surface composition. A radiometer will take Martian temperatures, while a radar instrument will search for underground water.

Beyond Mars

Now that the first great rush of solar system investigations has expanded our view of the planets, moons, and rings, upcoming missions are focused more on visiting some of the places that have only gotten passing attention – like comets and asteroids. No longer are these bits of rock and ice classified as merely solar system flotsam. They've come into their own as probes of conditions in the early solar system – clues to the conditions in which our Sun and planets formed. They are at once the key to understanding our ancient past as well as a new frontier for exploration.

Two missions have already been launched. The Stardust mission (discussed in chapter 2) is on its way to a rendezvous with Comet Wild, and already encountered

Figure 7.5. The Deep Impact mission will release an impactor to collide with Comet 9P/Tempel 1 in July 2005. By studying the material thrown out from the impact, and mapping the crater, the science instruments will return valuable data on the ices, gases, and dust that combined to make up the comet.

the asteroid Annefrank in November 2002. In January 2002, the Japanese MUSES-C probe began its multi-year mission to dock with the asteroid 1989ML in 2004, conduct a 3-month mission which includes picking up some samples, and then return to Earth in 2007. The Deep Impact mission is scheduled for a January 2004 launch to rendezvous with the comet 9P/Tempel 1. It will send an impactor to gouge out a crater on the comet, and study the material blasted out by the impact.

Deep Impact has as one of its components the Small Telescope Science Program, which brings together technically proficient amateur astronomers, professional astronomers with discretionary telescope time, and private observatories. Jointly they will gather valuable ground-based optical data on Comet Tempel 1 throughout the mission. These ongoing observations will supplement the data taken by the mission's science team, who will use the observations to determine the rotation rate of the comet's nucleus and its dust production rate. All of this information is crucial as the science team builds a model of the cometary environment so that they can accurately pinpoint the position and shape of the comet for the impact and imaging portions of the mission.

Perhaps one of the most exciting comet and asteroid missions is the European Space Agency's Rosetta Mission. It was named for the famous Rosetta Stone – the carved rock that yielded clues to the interpretation of ancient hieroglyphs. The original target was comet 67P/Churyumov-Gerasimenko, but the mission was postponed because a test launch failed. The mission plans include flybys of asteroids, flying alongside a comet for several months, and sending a lander spacecraft onto its surface. If the lander survives the touchdown, its sensing instruments will return information about the chemical makeup and the ever-changing conditions on the surface of a comet as it moves in its orbit.

Figure 7.6. The Rosetta mission is an ambitious project that will fly out alongside a comet and put a lander on its surface.

Exploring the Pluto/Kuiper Belt frontier

Just as every continent on Earth once had its untamed frontiers, the solar system has its unexplored regions. The last great mysterious deeps beyond the planets lie in the Kuiper Belt. This collection of objects floats out past the orbit of Neptune as part of a larger reservoir of icy objects including the Oort Cloud. As we discussed in chapter 3, many cometary nuclei are warehoused out in the Oort Cloud, just waiting for their chance at a headlong fling inward toward the warmth of the inner solar system. But the Kuiper Belt contains many larger icy bodies of interest.

A mission to the Kuiper Belt and Pluto has been on the planning boards for many years, but only in recent years has the urgency to get out there and explore reached a fever pitch. The reason? Pluto is currently passing out of its closest approach to the Sun. During its lengthy "summer" season, the planet warmed up enough to maintain a thin atmosphere and its wispy gas layer contains precious clues to the composition of ices that make up Pluto and Charon. There is the danger that Pluto's thin atmosphere will soon collapse and freeze onto the surface, and scientists will have to wait at least another century and a half before Pluto's blanket of gases will once again rise to become an "atmosphere." As the pair of worlds move farther from the Sun over the next decade, mission planners will lose favorable planetary alignments that would allow them to use Jupiter's gravitational field as a sort of slingshot to speed the spacecraft on its way to Pluto.

NASA, the Southwest Research Institute, and Johns Hopkins University, along with scientists from around the world, are planning to send a robotic probe called New Horizons to the Kuiper Belt for the express purpose of studying Pluto and its sister objects. It will carry miniature cameras, a radio science experiment, ultraviolet and infrared spectrometers, and space plasma experiments. The goal will be to map the surfaces of Pluto and Charon, and examine Pluto's collapsing atmosphere. After it completes its mission at Pluto, the spacecraft will turn its attention to other nearby Kuiper Belt objects. It will not be the most distant spacecraft humans have ever sent – that honor currently belongs to Pioneer 10 and the Voyager probes, but it will be the first time we will have a high-resolution look at the most distant objects in our solar system.

Are there other Earths out there?

Beyond the planets, astronomy missions of the next decade will focus on digging even more information out of the darkest, deepest places in the cosmos. Now that we know how and where to look for star birth, for example, the next logical step is to search out the planets that form along with stars. At the current time, the count of "hot Jupiters" – those objects that appear to be gas giants in orbit around other stars – stands at more than 100. However, for a variety of reasons, these aren't places where life is likely to spring up. Astronomers want to find places more like Earth, worlds where life might exist.

The 2006 Kepler mission is being designed to trail Earth in orbit around the Sun and sweep its gaze through the solar neighborhood looking for terrestrial-size and larger planets in or near the habitable zones of nearby stars. The habitable zone is roughly defined as that distance from a star where liquid water can exist on a planet's surface. As the spacecraft's sensitive instruments discover these planets, they'll be cataloged by their size, mass, density, and length of time they take to go

Figure 7.8. The Kepler mission to search for extrasolar planets will launch around 2006 and spend 4 years looking at nearby stars for evidence of other worlds like Earth.

around their parent stars. How will the mission find these places? It's well known that planets near stars are incredibly hard to see – in fact, in the case of worlds smaller than Jupiter-size, it's nearly impossible to spot them. However, when a smaller body (say a planet about the size of the Earth) passes in front of a star, it blocks out some of the star's light. To a distant observer with a very sensitive instrument, the star appears to dim for a brief time. Once it sorts out the stars that vary intrinsically (i.e., not because they have planets around them), the orbit and size of any planets can be calculated by studying the stellar transits. The Kepler instrument is essentially a very sensitive 0.95-m-wide photometer that can study a 105-square-degree starfield to monitor 100 000 main sequence stars at a time.

New observatories in space

The next major gamma-ray observatory to follow the 2002 International Gamma Ray Astrophysics Laboratory into orbit will be the Gamma-Ray Large Area Space Telescope (GLAST). With its 2006 launch, high-energy astrophysics will have a new look at objects radiating at the hottest temperatures in the universe. The mission's objectives are to determine the engines that power active galactic nuclei, pulsars, and supernova remnants. Beyond specific objects, the spacecraft will map the gamma-ray sky, concentrating on interstellar emissions from the Milky Way, gamma-ray bursters, and transients. And, as with so many other probes and observatories being planned for twenty-first-century astronomy, GLAST will make observations that may help solve the twin mysteries of dark matter and the activities that took place in the earliest epochs of the universe.

Figure 7.9. The James Webb Space Telescope will go into a Lagrangian point orbit sometime after 2010. It will replace the Hubble Space Telescope.

Joining in the search for the dimmest, most distant, and oldest objects in the universe will be the James Webb Space Telescope – the long-planned successor to the Hubble Space Telescope. If all goes well, the Webb telescope will be ready to take its place and focus its 6-m segmented active optical assembly and mostly infrared-optimized payload of instruments on galaxies so distant that their light has been stretched into the infrared wavelengths. The Webb Space Telescope is being called a "first light" machine because it will give astronomers a first-ever look at the universe as it appeared when it was a small fraction of its current age and size.

Although mission planners want to use this remarkable observatory to peer into the most distant reaches of the cosmos, it will have the sensitivity to study the oldest, dimmest stars in the Milky Way, the coolest regions of star-birth nurseries, and even the ability to take up the photometric detection of planets around other stars. Unlike HST, the Webb telescope will not be placed in low Earth orbit. It will take up a position at an orbital location called a Lagrangian point some 1.5 million km away.

Ground-based ventures

Ground-based astronomy is looking up in a big way, with some huge projects on the drawing boards. One of the largest is the eloquently named OWL project, which stands for OverWhelmingly Large telescope. This behemoth is in the planning stages, and if all goes well could open its 100-m optical and infrared eye on the sky as early as the year 2012. At present, the European Southern

Figure 7.10. The Overwhelming Large (OWL) telescope will aim a 100-m eye on the sky some time in the next two decades.

Observatory is doing a concept study. If it is built as planned, astronomers will be able to undertake long-term studies of star and planetary system formation. Its exoplanet studies will follow in the footsteps of today's searches, and those done by the Kepler mission. Beyond our galaxy, the OWL will be able to extend its gaze into the deepest reaches of the universe to search out the births of the earliest stars and galaxies.

Long before OWL is built, the next generation of large area radio telescope arrays should be well into their operational careers. The Atacama Large Millimeter Array (ALMA) has already built some prototype dishes in preparation for the full-scale construction of at least 64 12-m antennas on the Llano de Chajnator, nearly 5000 m up in the Andes Mountains of Chile. Among its targets will be heat emissions from gas, dust, and solid bodies in the universe. Right now, these emissions can be only studied from space at low resolution.

Another ambitious ground-based radio installation, the Low-Frequency Array (LOFAR), is planned to begin science observations in 2006. The 26 000-antenna array is being designed to scan the skies at the radio frequencies between 10 and 250 MHz, in an effort to gain new understanding of anything that radiates in that range – including cosmic ray bursts, merging galaxies, radio galaxies, and quasars. Theoretically, the array could be used to search the early universe and map the fluctuations in the cosmic microwave background. In the solar system the planet Jupiter, coronal mass ejections, and even the Earth's ionosphere could be fair game for LOFAR's radio receivers.

One of the most hotly studied areas of astronomy is the search for dark matter. As we discussed in chapter 6, the nature of this unseen but potent "stuff" is still largely unknown. Obviously it's dark, or at least so dim that none of our current instruments can directly see it. But we can infer its presence by its gravitational effects. What if we had a telescope dedicated solely to the search for dark matter?

Figure 7.12. A set of receiving stations in the Low-Frequency Array, which could begin initial operation in 2006. If all goes as planned, this extensive radio astronomy array will serve a number of student needs as well as projects from the professional community.

There are plans circulating for just such an instrument. This Large Synoptic Survey Telescope and its mission have been expanded from the search for non-luminous matter to include the search for near-Earth asteroids. LSST will be a ground-based 8.4-m, three-mirror design that will cover a whopping 260 square degree field of view in the sky. Linked to fast computers, the data from this installation would provide high-quality maps of very faint objects both near and far. It will be one of the largest surveys ever to look directly for the dark matter of the universe.

Another approach to astronomy research: Data mining among archived treasures

Today's astronomy researchers are heirs to an incredibly huge archive of past observational data. As we said in the introduction to this book, it's like a datastorm pouring out of the sky and onto the tape drives, CD-ROMs and other storage devices of the astronomy community. Because of the rapid advance in telescope and detector technology used on the ground and in space-based observing systems, astronomers have seen an incredible increase in both data quantity and quality. When you add in the torrent of information coming from automated sky surveys, it's easy to see how the volume of sky data has reached the multiple-terabyte level. And it just keeps coming. Much of that information is available in computerized form for nearly instant study by anyone who has the time and capability to do the work. Older data from such projects as the Voyager missions or ground-based observing sessions are being brought online as well.

Most of the world's facilities have some sort of data-sharing capability, and a good number of them also operate public outreach Web sites to make particularly

Figure 7.13. The Trifid Nebula.

interesting results available to the general public. Among the leaders in this "instant public access" of results are the Space Telescope Science Institute, the Jet Propulsion Laboratory, the National Optical Astronomy Observatories, the Chandra X-ray Observatory, the European Southern Observatory, the European Space Agency, and NASA. Any Web-savvy surfer can easily find enough information about astronomical objects in a few short minutes of search-engine browsing to satiate even the most voracious space-hound. In fact, this book would have been extremely difficult to create without some sort of data-mining capability, and advances in mailing lists and intelligent Web page design.

The massive libraries of raw data useful to astronomers dwarf the large volume of public-interest information. The existence of these observational data has posed problems in storage and retrieval of data for the end-users. Getting access to the data has also made things very challenging for astronomers who rely on past observations to plan their research and help analyze their newly received data. In recent years, scientists have proposed an International Virtual Observatory Archive. The idea behind this effort, undertaken largely by the United States National Virtual Observatory and the ASTROGRID (a collaborative research effort in the United Kingdom), is to create a so-called "front end" Web site that will link all the archived data sets from the various space missions, ground-based observatories, and sky surveys. A researcher from anywhere in the world would be able to log into this site, use commonly available search tools to comb the databanks for information pertaining to ongoing projects, and then use the data in his or her own research. In addition, the planners foresee the Virtual Observatory as a way for amateurs, graduate students, and the interested public to connect with astronomy data in a form they can use in their own research, education, or leisure pursuits.

Beyond the horizon

When we began this adventure, we told you that the cosmos dangles some incredibly beautiful things in front of us, coupled with some seemingly impassable distances. Then we showed you all those things out in the universe. The good news is that no matter how advanced our exploratory techniques become, there will always be another frontier to explore and discoveries to make. What is most important is for all of us to keep telling the next generation to keep exploring and learning.

The two young women who opened this final chapter are the future of astronomy. Perhaps you know someone like these budding astronomers. Perhaps you *are* someone like them. If so, here's a challenge for you. Gaze at the image of the Trifid Nebula presented here. We're not going to tell you about the technical details of this star-birth region. You can always go back and read about star birth in chapter 4, or look up the particulars of the image on the Gemini Web site that we have listed on page 207.

This challenge is really an exercise to populate your intellect with visions of the cosmos worth exploring. Think about what the image says to you. Where does it take you? How does it make you feel? The answers you come up with may surprise you, but then again, the universe has been doing that all along.

Glossary

In addition to the usual, short entries, this section also includes extended discussions of physical concepts helpful in understanding some of the research discussed in this book. **Bold** type denotes other entries in the glossary.

absolute temperature scale

In the physical sciences, temperatures are related to the motions of atoms and molecules in gases. The temperature scale used is the Kelvin scale, which has the same size degrees as the centigrade scale. On the Kelvin scale, the freezing point of water is 273 kelvin (K), the boiling point of water is 373 K, and the surface of the Sun (*effective temperature*) is 5750 K. The temperatures in this book are all given in kelvin.

absolute zero

A hypothetical concept to describe the temperature, 0 K, at which the motions of atoms and molecules would stop.

absorption line

When a continuous spectrum is viewed through a cooler, low pressure gas, such as the interstellar gas, dark lines called absorption lines appear. The specific lines depend on the composition of the gas that is absorbing the light and causing the lines.

active galaxy

A galaxy with a bright central region and often with enhanced x-ray, ultraviolet, and radio emission. Active galaxies include the Seyfert galaxies and **quasars**. The activity probably results from a massive, central **black hole**.

albedo

The amount of solar energy falling on a surface that is then reflected back to space. It is usually expressed as a fraction between 0 and 1. Earth's albedo, for example, is 0.39, meaning that it reflects 39 percent of the light back into space.

angles

Angular measure is frequently used in astronomy to specify positions on the sky and to specify apparent size. The units are degrees, minutes, and seconds. An entire circle contains 360 degrees; a right angle contains 90 degrees. Each degree contains 60 minutes and, in turn, each minute contains 60 seconds. Note that the minutes and seconds described here are not the same as minutes and seconds of time.

arcsecond

A commonly used, very small, angular measure in astronomy. Roughly speaking, it is one part in 200 000. To illustrate the size, a US quarter coin is about 2.5 cm across. If placed at a distance of 200 000 quarters or about 500 000 cm (5 km), it subtends an arc second. If we are talking about 0.1 arc second, it is about the angular size of a quarter 50 km distant.

asteroids	Sometimes called 'minor planets'. These are small, rocky bodies orbiting the Sun primarily between the orbits of Mars and Jupiter, the region of the "asteroid belt." Diameters range from the smallest detectable (at a few tenths of a kilometer) up to the largest asteroid, Ceres, at 950 km.
astrometry	The measurement of precise positions and motions, usually of stars.
astronomical unit (AU)	The average distance between the Earth and the Sun, i.e., about 150 million km.
atomic number	The number of protons in the nucleus of an atom (and hence the number of electrons in a normal, unionized, atom).
atomic weight	Approximately the number of protons and neutrons. For example, a helium atom, with two protons, two neutrons, has an atomic weight of 4.003.
baryons	A class of elementary particles that includes the **proton** and the **neutron**.
Big Bang	The postulated beginning of our universe when all the matter and radiation emerged from a point.
Big Crunch	If the universe should stop expanding and fall back on itself, it eventually could concentrate all the matter and radiation back into a point. This event is called the Big Crunch.
binary stars	Two stars in orbit around each other. If the separation can be resolved in the telescope, they are called visual binaries. If the double nature is revealed by motions detectable via the **Doppler effect**, they are called spectroscopic binaries. Most stars are believed to be members of binary or multiple systems.
black body	An ideal body which absorbs all radiation incident on it. The emission from a black body depends only on its temperature.
black hole	When massive stars collapse at the end of their lives, their mass is concentrated at a single point. The effective size of the star is called the **event horizon**, the boundary of a region with a gravitational field so strong that nothing can escape from it, not even light. Hence, the name "black hole." The lower limit for the mass of a black hole is about 3 M_S. Of course, this is the mass of the final core, not the original mass of the star on the **main sequence**, which can be much larger. Black holes with other origins are possible. So-called "mini black holes" may have formed early in the universe and "massive" black holes, with masses of millions to billions of solar masses, are probably located at the centers of galaxies.
blast wave	A traveling **shock** produced by an explosion.
carbon cycle	Nuclear reactions that convert four hydrogen nuclei into one helium nucleus with a release of energy using carbon nuclei as a catalyst. The carbon cycle occurs at higher temperatures than the **proton–proton chain**.

catalog numbers	Many astronomical objects in this book are designated by their catalog number. For example, M31 means the 31st object in Messier's catalog and NGC 188 simply means the 188th object in the New General Catalog. Others found in this book are Arp numbers, CL numbers, Henize numbers, Markarian numbers, Melnick numbers, PKS numbers, and 3C numbers. These are from specific catalogs or lists, often named after the astronomer compiling them. While we give the numbers for completeness, they are not essential for understanding their role in science.
Cepheid variables	Very luminous, pulsating stars that are important distance indicators. Their brightness pulsates regularly over a length of time called a *period*, and their **intrinsic brightness** can be determined from the period of light variation.
Chandrasekhar limit	The maximum mass of a **white dwarf** star or approximately 1.4 M_S.
cluster parallax	A distance determination method applicable to clusters with measurable motions toward a convergent point (where all the motions would intersect on the sky) and radial velocities determined by the **Doppler effect**. The distance to the Hyades **open cluster** has been determined using this method.
clusters of galaxies	Distinct groups of galaxies containing from 10 galaxies to thousands of galaxies.
clusters of stars	See **globular clusters** and **open clusters**.
comets	Solar system bodies consisting of an icy nucleus. If the comet approaches the Sun, the nucleus is heated, and the cometary ices sublimate to form a cloud of gas and dust called a coma. Sometimes the gas and dust will stream out away from the comet, pushed by the **solar wind** (for the gas) or the Sun's **radiation pressure** (for the dust) into a long tail.
constellation	One of the arbitrary groupings of stars in the sky, of which there are more than 80, which people imagine to look like objects, such as Orion the hunter in the winter sky (northern hemisphere). The brightest stars in each constellation are designated in order of brightness by a Greek letter in alphabetical order; for example, Alpha Orionis, Beta Orionis, etc.
continuous spectrum	An incandescent gas under high pressure (such as the sub-surface layers of the Sun) emits a continuous spectrum, i.e., emission at all wavelengths over a wide range of the electromagnetic spectrum.
cosmic abundances	A standard tabulation of the relative abundances of the elements in the universe compiled from solar, meteorite, stellar, nebular, and Earth-crust data.
cosmic microwave background (CMB) radiation	The nearly isotropic surviving radiation from the **primordial fireball** when the universe was 300 000 to 400 000 years old. It has a **black body** temperature of 2.73 K.

cosmological constant	A form of **dark energy** introduced by Einstein in 1917. The repulsive force remains constant with time.
cosmology	The study of the entire universe considered on a very large scale; this includes the origin, structure and evolution of the universe.
critical density	For a matter-dominated universe, the density of the universe that would just stop the expansion of the universe after a very long time is called ρ_0. For a **Hubble constant** of 70 km per second per **megaparsec**, this amounts to 0.9×10^{-29} g per cubic cm, a very small number. If this were all hydrogen atoms, their density would be 0.5×10^{-5} atoms per cubic cm. A more convenient quantity is Ω_0, which is the ratio of the actual density to the critical density. For an actual density equal to the critical density, Ω_0 is 1. For Ω_0 higher than 1, the expansion stops and becomes a contraction, ultimately leading to the **Big Crunch**. For Ω_0 lower than 1, the universe expands forever.
	For an energy-dominated universe, Ω_0 becomes the sum of Ω_{matter} (total matter) and Ω_Λ (dark energy). Then, Ω_0 determines the geometry of the universe. For Ω_0 less than 1, the universe is open. For Ω_0 equal to 1, the universe is flat. For Ω_0 greater than 1, the universe is closed. The current evidence indicates that we are in a **flat universe**.
dark energy	A general term for the ubiquitous form of energy in the universe that produces repulsion. See **cosmological constant** and **quintessence**.
dark matter	Unobserved, and hence "dark," matter in the universe. This matter could be of a straightforward nature, such as very sub-luminous stars, and/or it could be exotic sub-atomic particles. The evidence for large amounts of dark matter in the universe is extensive.
deceleration parameter	A measure of the rate at which the expansion of the universe may be slowing down because of the gravitational attraction of its own mass.
deconvolution	The sharpening of a degraded image (or spectrum), usually by computer processing. If the degradation can be accurately determined, for example by determining the degraded appearance of a star which should be nearly a point source, it can be substantially removed.
deuterium	A heavy hydrogen atom containing one proton and one neutron in its nucleus.
Doppler effect	A change in the wavelength of light caused by a relative motion of a source and its observer. This change in wavelength can be understood by thinking of light as a wave. If the source approaches the observer, more waves per second are seen and the frequency is therefore raised, and the wavelength is shortened; this is referred to as a blue-shift – a shift towards the "blue" end of the spectrum. When the source is moving away from the observer, the opposite occurs; this is referred to as redshift. We hear the same effect in the sound of a siren; the pitch (frequency)

is higher (corresponding to a shorter wavelength) when the vehicle approaches us and the pitch is lower (corresponding to longer wavelength) when the vehicle is going away from us, so as it passes, a distinct change in the pitch of the siren is noticed.

effective temperature The temperature that an ideal body, called a **black body**, would have if it were to emit the same amount of energy per unit area. The term is usually applied to stars; for example, the Sun's effective temperature is about 5750 K.

electromagnetic spectrum (EMS) The entire spectrum of all forms of electromagnetic radiation, including light, from gamma rays to radio waves.

electron The light, negatively charged particle in the atom.

emission line Incandescent gases at low pressure, such as a gaseous nebula, produce a spectrum composed of individual bright lines called emission lines. The concept of "line" means that the emission occurs at a specific wavelength or narrow range of wavelengths, a situation in contrast to the **continuous spectrum**. The specific lines depend on the composition of the gas. See also **absorption line**.

event horizon The surface, or effective boundary of a **black hole**. No material or light can escape from within the even horizon. The radius is approximately 3 km times the mass of the black hole (given in solar units).

fission The break-up of atomic nuclei into lighter nuclei.

flat universe A universe where the geometry is Euclidean. This means that parallel lines remain parallel when extended into the distance and that the sum of the interior angles of a triangle is 180°. This geometry applies to our universe. See **critical density**.

fusion The merging of atomic nuclei into heavier nuclei.

galaxy One of the very large celestial objects consisting of stars (up to roughly a trillion, 10^{12}) and often vast quantities of dust and gas. The Sun is in the **Milky Way galaxy**. External galaxies are often described by their appearance, such as spirals, barred spirals, ellipticals, and irregulars.

general relativity Albert Einstein's generalization of **Newtonian mechanics**, which is used in cosmological applications.

giant A star roughly 100 times the Sun's intrinsic brightness with a radius of roughly 100 times the Sun's radius.

globular clusters Tightly packed, spherical groupings of old stars in the **Milky Way galaxy**. Globular clusters are also observed in other galaxies.

gravitational lensing	The general relativistic effect whereby the enormous mass of celestial objects changes the path of light passing by. The effect is to focus the light and thus the object causing the effect is, in essence, a gravitational lens.
gravity	The attraction of all bodies in the universe for all other bodies. Two bodies attract each other with a force proportional to the product of their masses and inversely proportional to the square of the distance between them. For example, the solar gravitational force acting on the Earth would be one-quarter the present value if the Earth were twice as far from the Sun as it is now.
Hertzsprung–Russell (H–R) diagram	A display of stellar properties using a plot of **effective temperature** (or surrogates such as color or spectral type) versus **luminosity** (or surrogates such as **intrinsic brightness**).
Hubble constant (H_0)	The constant of proportionality between the recession speeds of galaxies and their distances from each other. Current estimated values range between 50 and 100 km per second per **megaparsec**. The preferred value is close to 70 km per second per **megaparsec**.
inflation	A hypothetical period of extraordinarily rapid expansion early in the universe, from roughly 10^{-34} to 10^{-30} seconds after the **Big Bang**. This expansion may be necessary to explain properties of the universe, such as the existence of galaxies. A consequence of inflation is a flat geometry for the universe; see **flat universe**.
in situ	The Latin term meaning "in position," often used by scientists talking about explorations at a planet, or on its surface.
interstellar medium	The gas and dust located between the stars in the Milky Way galaxy.
intrinsic brightness	The brightness of an object, such as a star, that is independent of distance. This brightness can either refer to light in a specific color or to all the light, when it is the same as the **luminosity**.
ion	An atom or molecule that has lost one or more **electrons** and thus has an electrical charge.
isotope	Atoms of the same chemical element, but with different numbers of neutrons in the nucleus.
kiloparsec	1000 **parsecs** or 3.1×10^{16} km.
Kirchhoff's laws	Experimentally determined ideas based on the absorption and emission of light. See **absorption line**, **continuous spectrum**, and **emission line**.
Lagrangian point	One of five points in the orbital plane of two massive objects circling a common center of gravity where a smaller body of negligible mass is in equilibrium. For spacecraft, two of these points in Earth's orbital plane are where a satellite can

remain in stable orbit. The other three are unstable, but can be used with the aid of orbital thrusting by the spacecraft.

light-travel time	The time it takes for light, traveling at about 300 000 km per second, to travel a certain distance.
light-year	The distance that light travels in 1 year at about 300 000 km per second, i.e., 9.5×10^{12} km.
look-back time	Because light travels through space at a constant speed (300 000 km per second), it takes a finite time to travel from distant objects. Hence we "see" distant objects at a point in time in their past; this point in time is the look-back time. See **light-travel time**.
luminosity	The total light output of a star over all wavelengths. Because this quantity is independent of distance, it is an **intrinsic brightness**.
Lyman-alpha	The principal light-absorbing or light-emitting energy transition of the hydrogen atom occurring at about 121.6 **nanometers**. Often abbreviated as Lyman-α.
Magellanic Clouds	Two irregular galaxies, the Large Magellanic Cloud (LMC) and the Small Magellanic Cloud (SMC), easily visible to the unaided eye in the southern hemisphere. They are named after explorer Ferdinand Magellan.
magnitudes	Stellar brightness is often described in terms of a historical system that seems confusing. The magnitude system is somewhat like standings in sports leagues where the best (and brightest) are first, the lesser (or fainter) are second, third, etc. Historically, the brightest stars (collectively) in the sky are first magnitude and the faintest stars visible to the naked eye are sixth magnitude. The modern definition is that stars differing by a factor of 100 in brightness differ by 5 magnitudes. This works out to a factor of 2.512 in brightness for one magnitude, and the fainter stars have the larger magnitude. The magnitudes as observed for stars on the sky are called apparent magnitudes. If the magnitude is referred to the standard distance of 10 **parsecs** (and thus is indicative of the star's **intrinsic brightness**), it is called an absolute magnitude. Finally, the magnitude can refer to all the light from a star or to the brightness of light measured through a specific filter, such as a blue filter.
main sequence	The principal band of stars on the **Hertzsprung–Russell diagram**. Most stars appear on the main sequence after nuclear burning of hydrogen has begun, and they spend most of their lives there.
megaparsec	One million **parsecs** or 3.1×10^{19} km.
Milky Way galaxy	Our own galaxy, consisting of about 100 billion stars plus gas and dust. The galaxy is disk-shaped; when we see the Milky Way in the sky, we are looking at the disk edge on.

nanometer	The unit of wavelength commonly used in astronomy and abbreviated nm. It is 10^{-9} meters. Blue light, for example, has a wavelength around 470 nm.
nebula	A cloud of gas and dust in a galaxy, such as the Orion Nebula. Before the early part of the twentieth century, this term was used to describe any hazy patch in the sky.
neutrino	A stable particle with no charge and no mass that is produced in nuclear reactions. The neutrino interacts very weakly with all other particles.
neutron	The heavy, electrically neutral elementary particle in the nucleus of an atom.
neutron star	A very dense star composed of **neutrons** and having a diameter of about 30 km. The pressure in the star is so great that **electrons**, which usually orbit the nucleus of atoms, are pressed into the nucleus where the protons and electrons merge to form neutrons. The mass of neutron stars falls between 1.4 and 3.0 M_S. Smaller stars become **white dwarfs** while larger stars become **black holes**. Neutron stars are believed to result from **supernovae**. **Pulsars** are believed to be rapidly rotating neutron stars.
Newtonian mechanics	Theory (or equations) for calculating the positions of celestial bodies such as planets and stars. This theory is based on Isaac Newton's laws of motion and gravity.
nucleosynthesis	The creation of the chemical elements by way of nuclear reactions. The formation of the light elements – hydrogen, **deuterium** (a form of hydrogen), helium, and lithium – occurred during the Big Bang. Other elements are created in stellar interiors and during supernova explosions, where temperatures and pressures are high enough to facilitate the necessary energetic collisions.
open cluster	A stable grouping of young stars in the **Milky Way galaxy**. Open clusters present in other galaxies are good indicators of spiral arms.
parallax	The apparent shift in position of a nearby object projected on a background when viewed from different positions. Simply holding a finger at arm's length and viewing it with only one eye and then the other illustrates the effect. In astronomy, the parallax is the shift in position of a star when viewed from positions separated by 1 **astronomical unit**. For nearby stars, an accurate parallax fixes their distances.
parsec	The distance, 3.1×10^{13} km, at which a star has an astronomical **parallax** of 1arcsecond. Used as a unit of measurement.
period	The length of time it takes for a body to go from a starting position or condition and return to that position or condition. Period can be used to describe such actions as the rotation of a planet on its axis or its orbit around a star; and the length of time it takes a star to vary from maximum to minimum brightness and back to maximum.

photon	A particle of electromagnetic radiation. Each photon carries an energy proportional to its frequency.
planetary nebula	A **nebula** which resembled a planetary disk in early telescopes. Now known to be an expanding gas cloud surrounding a hot star.
planets	The major solar system bodies in orbit around the Sun. In order of increasing distance from the Sun, they are: Mercury, Venus, Earth, Mars, Jupiter, Saturn, Uranus, Neptune, and Pluto.
populations	A concept used to describe different kinds of stars, which are usually distinguished by age and abundance of heavy elements. Population I stars are young and have a high abundance of heavy elements, while Population II stars are old and have a low abundance of heavy elements.
powers of ten notation	A mathematical shorthand used to conveniently record very large or very small numbers. For example, 1000 is $10 \times 10 \times 10$ or 10^3. The number 2000 is 2×10^3. One hundredth or 0.01 is just $1 \div (10 \times 10)$ or 10^{-2}. Basically, the power of ten simply locates the decimal point. This notation also allows easy multiplication and division. The powers add when multiplying and subtract when dividing. For example, 100×1000 is $10^2 \times 10^3$ or 10^{2+3} or 10^5.
primordial fireball	The state of the universe that existed after the **Big Bang** consisting of energetic elementary particles and radiation. The **cosmic microwave background radiation** is a remnant of this era.
proton	The heavy, positively charged elementary particle in the nucleus of the atom.
proton–proton chain	Nuclear reactions that convert four hydrogen nuclei into one helium nucleus with a release of energy. This chain occurs at lower temperatures than the **carbon cycle** and is the principal energy source for the Sun.
proto-planets	Planets at an early stage of formation from clumps of gas and dust.
proto-stars	Stars at an early stage of formation from interstellar clouds of gas and dust.
pulsar	A rapidly rotating **neutron star** which emits characteristic pulses of radio radiation and visible light.
quantum mechanics	The physical laws that describe the motions of electrons in atoms. These laws are radically different from **Newtonian mechanics** in that only certain electron orbits or energies are allowed. Thus, atoms emit or absorb light in discrete amounts called quanta. See **absorption line, continuous spectrum**, and **emission line**.
quasar	Also called quasi-stellar objects or QSOs, quasars are star-like objects which produce emission lines. Their redshifts can be large and their brightness varies.

They are believed to be objects a little larger than the solar system that have an intrinsic brightness some 100 times that of bright galaxies.

quintessence A form of **dark energy** that varies with space and time. Literally, it is the "fifth element," by analogy with the elements of ancient Greek philosophy.

radiation pressure The tiny force exerted by photons when they bounce off small dust particles or when they are absorbed by atoms. The force is so small that it has no effect on the large objects encountered in everyday life, but it can be important in astronomy.

redshift The redshift is a shift toward the red end of the spectrum seen in the wavelengths of light put out by an object. Astronomers often quote the distance of a galaxy in terms of its redshift and assign it a quantity z. This shifting of light is caused by the Doppler effect (the cosmological effect of the expansion of the universe) or a gravitational redshift that occurs as the photons emitting the original wavelengths move through a gravitational field. The value of z gives the redshift in units of the original wavelength. If a spectral line is originally at 200 nm, but was measured at 800 nm, then there has been a threefold shift in the line. The object's redshift or $z = 3$. An object with $z = 2$ has a spectrum shifted 400 nm and its wavelengths are now measured at 600 nm. A $z = 1$ produces a new wavelength of 400 nm while $z = 0$ is no shift at all.

resolution See **spatial resolution**, **spectroscopy**, and **temporal resolution**.

shock A sharp change in the properties (density, pressure, temperature) of a gas.

signal-to-noise Measurements of light in astronomy are always composed of two parts. The first is the signal, the part produced by the pure light from the target. The second is the noise, the part that comes from other sources, such as the telescope, the detectors, etc. The signal-to-noise ratio indicates the quality of the measurement.

solar nebula The cloud of gas and dust from which the solar system, including the Sun, formed.

solar wind The low-density gas consisting primarily of **protons** and **electrons** flowing away from the Sun at about 500 km per second.

spatial resolution The measure of the ability of a telescope and camera to separate objects clearly. Roughly, the smallest detail that can be seen in an image. Compare **temporal resolution**.

spectral lines Because **electrons** in the cloud around the nucleus of an atom can have only a restricted number of specific energies, changes in energy produce **photons** of energy unique to that particular atom. This is the origin of spectral lines. These specific lines show that an element is present, and the strength of the line can be used to determine how abundant the element is. Shifts of the line in wavelength via the **Doppler effect** give the motion (speed) toward or away from the observer. The motion of atoms (with a speed depending on temperature) will cause a broadening of the line because some of the motions are away from and some are toward the observer. Thus, line widths can yield temperatures. Densities can be determined from the strengths of several different lines or from line widths. Finally, magnetic and electrical fields, if strong enough, can affect the the appearance of the lines. See *absorption line*, *emission line*, and *spectroscopy*.

spectral resolution

See *spectroscopy*.

spectroscopy

The analysis of light to determine the properties of the medium producing it and/or influencing it. In principle, spectroscopy can yield compositions, abundances, speed (via the *Doppler effect*), temperatures, densities, and, if they are strong enough, magnetic and electric fields. The rise of modern astrophysics clearly parallels the development of spectroscopy. To carry out spectroscopy, we need to disperse, or split up, the light into its component wavelengths or colors. This can be done by means of a prism, but in modern spectrographs, the dispersive optical element is a diffraction grating, which consists of a system of precisely aligned, narrowly spaced grooves. When carrying out spectroscopy, sufficient spectral resolution is needed to separate features in the spectrum. Numerically, the spectral resolution is given by the wavelength (λ) divided by the difference in wavelength to the closest wavelength that can be separated ($\Delta\lambda$). Thus, $\lambda/\Delta\lambda$ gives us a measure of the spectral resolution of instruments, such as the Hubble Space Telescope's spectrographs. The concept of spectral resolution is the spectral analog of *spatial resolution* that determines the smallest detail that can be seen in ordinary images.

spiral arm

The region in a spiral *galaxy* (such as the **Milky Way galaxy**) that contains concentrations of gas, dust, and young stars.

stellar wind

The general, steady flow of gas away from stars resulting in loss of mass. These range from the gentle **solar wind** to vigorous flows some 100 million times stronger in mass loss.

supergiant

A star with maximum intrinsic brightness and low density. The radius of a supergiant can be as large as 1000 times that of the Sun.

supernova

An immense stellar explosion which can increase a star's **intrinsic brightness** by as much as a billion times. For some types of supernovae, the explosion blows off a major fraction of the star to form an expanding gas cloud (such as the Crab Nebula). The remaining material forms a dense (compact) object such as a **neutron star** or a **black hole**. Type Ia supernovae (sometimes called **white dwarf** supernovae) occur when the mass of the white dwarf is pushed over the **Chandrasekhar limit** and the star is consumed in a thermonuclear explosion. Type Ia are intrinsically the brightest supernovae. They appear to be standard candles and they do not leave a compact remnant.

synchrotron radiation	The radiation emitted when accelerated charged particles are spiraling in a magnetic field.
temporal resolution	The measure of the ability of an optical system to clearly separate events in time. Roughly, the shortest time interval that can be determined between two different events. Compare **spatial resolution**.
vacuum energy	The lowest (but finite) energy of so-called empty space. It may be the physical basis of the **cosmological constant**.
white dwarf	An approximately Earth-sized star that does not have a source of nuclear energy in its interior. The star is supported by means of a form of pressure that arises when the densities at the star's interior are so high that the usual orbits of electrons in atoms around the nucleus cannot exist and the electrons are pushed much closer to the nucleus. A white dwarf can be supported in this way as long as its mass does not exceed about $1.4\ M_S$. For stars with greater masses, **neutron stars** or **black holes** are formed.
z	The symbol assigned to a **redshift** value.

Websites

The World Wide Web is a treasure trove of good astronomy and space science sites. The following pages are among the most publicly accessible sites and most lead to collections of other sites.

Amateur Astronomy Magazine
http://www.amateurastronomy.com/index.html

American Association of Variable Star Observers
http://www.aavso.org

American Astronomical Society
http://www.aas.org

Anglo-Australian Observatory
http://www.aao.gov.au/

The Astronomical League
http://www.astroleague.org/

Astronomical Society of the Pacific
http://www.astrosociety.org

Astronomy Now
http://www.astronomynow.com
(Britain's leading astronomy magazine)

Astronomy Picture of the Day
http://antwrp.gsfc.nasa.gov/apod/astropix.html

British Astronomical Association
http://www.britastro.org/index.html

Canadian Galactic Plane Survey
http://www.ras.ucalgary.ca/CGPS/

Chandra X-Ray Observatory Center
http://chandra.harvard.edu

Comet Observation Home Page
http://encke.jpl.nasa.gov

Earth from Space
http://earth.jsc.nasa.gov

European Southern Observatory
http://www.eso.org

European Space Agency
http://www.esa.org

Gemini Observatory
http://www.gemini.edu/

Malin Space Science Systems
http://www.msss.com

International Dark-Sky Association
http://www.darksky.org

National Aeronautics and Space Administration (NASA)
http://www.nasa.gov
(a portal to many different parts of NASA)

National Observatory of Japan (NOAJ)
http://www.naoj.org/
(Home of the Subaru Telescope)

National Oceanic and Atmospheric Administration (NOAA)
http://www.photolib.noaa.gov/collections.html

National Optical Astronomy Observatories (NOAO)
http://www.noao.org

National Radio Astronomy Observatory (NRAO)
http://www.nrao.org

Phil Plait's Bad Astronomy
http://www.badastronomy.com

Planetary Photojournal: NASA's Image Access Page
http://photojournal.jpl.nasa.gov

Sky and Telescope Magazine
http://skyandtelescope.com/

SkyNews
http://www.skynewsmagazine.com
(Canada's astronomy magazine)

Space Telescope Science Institute
http://oposite.stsci.edu
http://hubblesite.org

StarChild Learning Center for Young Astronomers
http://starchild.gsfc.nasa.gov/docs/StarChild/StarChild.html

Super-Kamiokande Neutrino Observatory
http://www-sk.icrr.u-tokyo.ac.jp/doc/sk/index.html
(part of Kamioka Observatory, Japan)

2-Micron All-Sky Survey (2MASS)
http://www.ipac.caltech.edu/2mass/

Books

The following list of books is not meant to be an exhaustive selection of reading but rather represents an interesting collection of reading material should you wish to delve more deeply into the cosmic visions we discussed here.

J. Bennett, M. Donohue, N. Schneider and M. Voit (1999) *The Cosmic Perspective*. Addison Wesley Longman. (textbook)

P. A. Charles and F. D. Seward (1995) *Exploring the X-Ray Universe*. Cambridge University Press. (textbook)

J. Kaler (1994) *Astronomy*. Harper Collins College Publishers. (textbook)

S. Kwok (2002) *Cosmic Butterflies*. Cambridge University Press. (general)

J. M. Pasachoff (2000) *A Field Guide to the Stars and Planets*. Houghton Mifflin. (general)

J. Beatty, C. C. Petersen and A. Chaikin (1999) *The New Solar System*, 4th edn. Cambridge University Press. (general)

T. Siegfried (2002) *Strange Matters*. Joseph Henry Press. (general)

Magazines

Astronomy Magazine, Kalmbach Publishers, Waukesha, Wisconsin (aimed at astronomy enthusiasts)

Sky and Telescope Magazine, Sky Publishing Corporation, Cambridge, Massachusetts (aimed at amateur astronomers)

Natural History, American Museum of Natural History, New York

Physics Today, American Institute of Physics (aimed at physicists and students of physics)

Science Magazine, American Association for the Advancement of Science, Washington, D.C.

Scientific American, Scientific American, Inc., New York

Chapter 2

Figure 2.1:   NASA JSC; NASA JPL; Mark C. Petersen; Carolyn Collins Petersen; Royal Swedish Academy of Sciences

2.2:   Gemini Observatory; McDonald Observatory, University of Texas at Austin

2.3:   Axel Mellinger; 2MASS; National Radio Astronomy Observatory; Extreme Ultraviolet Explorer (NASA); Röntgensatellit (ROSAT) (ESA); Compton Gamma-Ray Observatory (NASA)

2.4:   ESO

2.5:   Copyright 1977–2002, Anglo-Australian Observatory, photograph by David Malin

2.6, 2.8:   National Optical Astronomy Observatory/Association of Universities for Research in Astronomy/National Science Foundation

2.7:   Gemini Observatory; Subaru (National Observatory of Japan); James Clerk Maxwell Telescope, Mauna Kea Observatory, Hawaii

2.9:   Smithsonian Astrophysical Observatory

2.10:   National Astronomy and Ionospheric Center (NAIC) – Arecibo Observatory, a facility of the National Science Foundation; National Radio Astronomy Observatory; Ian Morrison, Jodrell Bank Observatory; Dave Finley, National Radio Astronomy Observatory

2.11:   California Institute of Technology; Kamioka Observatory, Institute for Cosmic Ray Research, University of Tokyo

2.12:   Gary Emerson

2.13:   NASA JSC

2.14, 2.15:   Chandra X-ray Observations Center

2.16, 2.18:   NASA

2.17, 2.19, 2.20:   NASA JPL, California Institute of Technology

2.19:   NASA JPL

Chapter 3

3.1:   NASA JPL

3.2, 3.14:   Image courtesy of Earth Sciences and Image Analysis Laboratory, NASA JSC

3.3, 3.12a, 3.15, 3.17:   STScI

3.4:   STScI; ESO

3.5:   SOHO; STScI

3.6:   Earth from Space, NASA JSC

3.7:   Department of Defense, Ballistic Missile Defense Organization; NASA JPL; Malin Space Science Systems

3.8:   Near-Earth Asteroid Rendezvous mission (NASA)

3.9:   NASA JPL

3.10:   Copyright 2002 by Fred Espenak, www.MrEclipse.com

3.11:   Courtesy Alan Treiman, Lunar and Planetary Laboratory

3.12b: University of New Mexico

3.13:   John T. Clarke, University of Michigan; NASA[211]

3.16:   ESO; National Radio Astronomy Observatory

5.5, 5.14:   National Observatory of Japan

5.10, 5.13, 5.17:   ESO

5.11, 5.12:   National Optical Astronomy Observatory/Association of Universities for Research in Astronomy/National Science Foundation

5.16:   Infrared Astronomy Satellite; Jayanne English (University of Manitoba) using data acquired by the Canadian Galactic Plane Survey National Research Council/Natural Sciences and Engineering Research Council (Canada) (NRC/NSERC) and produced with the support of Russ Taylor (University of Calgary)

5.18:   STScI; Tom Jarrett, IPAC California Institute of Technology

5.19, 5.20:   ESO; STScI

5.23:   Chandra X-Ray Observatory Center; National Radio Astronomy Observatory

5.30:   National Optical Astronomy Observatory/Association of Universities for Research in Astronomy/National Science Foundation; STScI; Chandra X-ray Observatory Center; STScI

Chapter 6

6.1:   John C. Brandt

6.2:   Wendy Freedman; G. Dindermann

6.3, 6.9, 6.13, 6.14a,b: STScI

6.4:   Cosmic Background Explorer (COBE) NASA

6.5:   2MASS

6.6:   Matthew Colless (Australian National University), and the 2 Degree Field 2DF Galaxy Redshift Survey

6.7:   Carolyn Collins Petersen (adapted from J. Ostriker)

6.8a,b: Hubble Space Telescope, A. Fillippenko, W. Li

6.10 Adam Riess *et al.* (NASA)

6.11a, b: Max Tegmark

6.12:   Courtesy Wilkinson Microwave Anisotropy Probe (NASA)

6.15, 6.16:   A. Tyson, Bell Labs, Lucent Technologies

6.17:   Kamionkowski is spelled with an "n" not a "u"

6.17 Based on materials supplied by E. Hivon and M. Kamionkowski, California Institute of Technology

Chapter 7

7.1:   Gemini Observatory

7.2:   ESA

7.3, 7.4, 7.5:   NASA JPL

7.6:   ESA

7.7:   NASA/Johns Hopkins University/Southwest Research Institute

7.8:   NASA/SETI Institute

7.9:   STScI

7.10, 7.11:   ESO

7.12:   Low-Frequency Array (LOFAR) Consortium; M.C. Petersen

7.13:   Gemini Observatory

Index

Page numbers in italics indicate illustrations.

214